BRIDGE

AND

TUNNEL CENTRES.

BY

JOHN B. McMASTER, C. E.

NEW YORK:
D. VAN NOSTRAND, PUBLISHER,
23 MURRAY AND 27 WARREN STREET.
1875.

PREFACE.

It is the purpose of the following essay to present in as brief a manner as the nature of the subject will allow, the rules and principles, the application and observance of which is of really vital importance in the planning and construction of Bridge and Tunnel Centring. It is not offered as a highly elaborated and exhaustive treatise on this branch of engineering, nor is it intended to furnish a variety of designs likely to be useful to the carpenter and bridge-builder ; it does not claim to be analytical ; it is purely practical.

Very much, therefore, of what, under other circumstances, might most fittingly have been introduced, has been carefully omitted, and nothing set down which does not bear directly on the subject in hand, and had not been verified, time and again, by actual experiment. It will be observed, for instance, that what may be termed the mathematics of the subject finds no place here. There are no mathematical demonstrations, no lengthy discussions of the various formulæ introduced ; they are simply set down as expressing established truths, the proof of their correctness in many cases suppressed, and the reader requested to accept them as true. In the form in which this essay first appeared, this was done to save space ; in the present form it has been strictly adhered to, because it is believed that those to whom the work will be of the most use, are precisely those who will be content to take as true the formulæ given, caring very little for the steps by which it has been reached.

In connection with the matter of estimating the load on a centre, four methods have been selected, either of which will give results close enough to the absolute truth for all practical purposes. The first, that of M. Couplet, is extremely simple, and if it errs at all, does so on the side of safety. The second or "graphical" is constantly growing in favor, and most deservedly so ; the third, that by calculus, disregards friction and give results greatly in excess of the truth, while the fourth, or trigonometrical, is perhaps the most exact of all, and admits the application of logarithms.

The remarks on the subject of uncentring are believed to be sufficiently extended, though the subject is one of great importance. The principles, however, to be observed in striking centres are quite few and simple, the observance being all that is necessary to secure success. The sand method cannot be too highly commended.

The remarks coming under the head of tunnel centres, have been limited to pointing out the essential difference between the centre proper for bridging, and that suitable for tunnelling, to calling attention to the peculiar variability of the strains, and to the care to be observed in guarding against the accidents so liable to produce injury to the ribs, and to offering a few practical suggestions as to economy. A few designs have also been added as illustrative of the principles laid down, and as affording examples of cheap and durable frames. The patent centre of Mr. Frazer is worthy of some attention.

JOHN B. McMASTER.

NEW YORK, November, 1875.

BRIDGE AND TUNNEL CENTRES.

In the construction of stone and brick arches, of whatever shape and span, and to whatever use applied, whether as supports for roadways or roofs of tunnels, there is nothing which requires more careful attention on the part of the constructing engineer, than the centres. Independent of the choice of material, of the exactness with which each stone is cut, and the care with which it is laid in place, the success of arches of great span, their settlement and ultimate stability depends essentially on the care given to the framing, setting up and striking of the centres. The slightest change in the shape of the frame caused by the shrinking of an ill-seasoned timber, or the yielding to compression of a badly proportioned brace, will assuredly be followed by a change in the curve of the intrados, which may possibly result in the ruin of the arch itself.

Well constructed centring, therefore, is indispensably necessary to a well constructed arch, and in the following papers it is our intention to offer a practical investigation of the principles which must be followed out in the planning and mechanical execution of all such centre frames ; to determine what strains must be withstood, at what point they act with most vigor, and by what combination of beams and by what system of bracing, the greatest strength and stiffness may be combined with the utmost lightness and the strictest economy of material.

BRIDGE CENTRES.

Of all classes of centres, the most complicated in structure is, beyond doubt, that of a large span stone bridge. Like a roof frame, it consists of a number of vertical pieces, placed in the direction of the span, from 5 to 7 ft. from centre to centre, and known as the *ribs*, upon which are placed horizontal pieces or *laggings*, and on these latter rest the voussoirs till the key stone course is driven and the arch becomes self-supporting.

THE FRAME in its turn is composed of *back pieces*, or short beams cut on the outer edge to the same curve as the intrados of the arch, a horizontal *tie beam*, and a number of *struts*, *ties* and *braces*, the arrangement, number and dimensions of which, will depend on the shape and span of the arch, and the number and position of the points of support. Whatever may be the span and curve of the arch, and the points of support afforded, experience has amply proved that the ribs should be polygonal in shape, with short sides; this shape being given by forming the back-pieces, on which rest the laggings, of two or more courses of planks, placed in the form of a polygon and firmly nailed together; the planks in each course abutting end to end by a joint in the direction of the radius of curvature of the arch, and breaking joints with those of the other course.

For light arches of moderate span, or indeed for heavy arches of wide span when firm intermediate points of support can be had between the abutments, the back pieces may be strengthened by struts or ties placed under them, well braced, and abutting against a horizontal tie beam. This beam spans the arch a little above the springing line, is bolted to the back-pieces at either side, thus preventing them from spreading laterally, and if well sustained by props from beneath, affords a firm support to the struts and braces of the rib. In by far the greater number of cases, however, where headway is required under the centring during the construction of the arch, as is the case with stone bridges spanning a river whose navigation cannot be impeded, or whose current is too swift and depth too great to give firm points of support to the props of the tie beam, it becomes necessary to do away with the latter, and supply its place by such an arrangement of beams as will transmit the strains received to points of support at the abutments. This latter class of centring is known as "*retroussée*" or "*cocket*," and requires a much more careful and elaborate arrangement of its parts than the former.

We have therefore two classes of bridge centres to deal with; one in which the frame is constructed without regard to headway beneath it, and is supported from firm points of support between the abutments, and one arranged to leave headway under the frame, and upheld by framed supports at the abutments.

Before attempting to determine the most advantageous arrangement of the pieces which must compose the frame, their number and the dimensions it is necessary to give them in order that they may offer a solid support to the arch stones, it is fitting to consider the effect of the load the ribs are expected to uphold, the strains it produces, the points where and the directions in which the strains act and their intensity.

THE STRAINS.

The strains to which centre frames are subjected arise solely from the pressure upon the back-pieces and laggings, due to the weight of the voussoirs laid upon them, and are therefore extremely variable, depending on the span and curve of the arch, and the thickness and weight per cubic foot of the voussoirs which press upon the centring. It is not, however, to be supposed that *all* the voussoirs from springing line to springing line *do* press upon the frames, this depending to a very great degree on the curve of the arch. If, for example, we take the case of a full centre arch and starting at the springing line on either side pass towards the crown, we shall find that for a considerable distance above the springing line the stones do not exert any pressure upon the ribs, but that, as soon as this point is passed, the pressure begins and increases rapidly, reaching its maximum intensity just before the keystone course is driven into place. When this is done the pressure is almost entirely removed, and were it not for the slowness of the mortar in drying, the frame work of the arch might be done away with.

And, here, I would mention that, although it is generally held that when the keying course is placed, the voussoirs, with the exception of a few courses at the crown, cease to press, I have found by the most careful experiments with large, well-framed models, that the thinnest Chinese paper when coated with black lead and placed under the blocks of arch stone, *could not* be *drawn out*, even when the arch was keyed, without considerable resistance.

Upon further examination it will be found that these voussoirs which lie near the springing line and exert no pressure upon the laggings and back-pieces, are all of them contained within the angle of repose ; that is to say, the voussoirs do not begin to press upon the centring until we reach one whose lower joint makes so great an angle with the horizon, that the stone is caused to slide along its bed under the action of gravitation. This angle for full centre arches has been fixed at from 28° to 30°, but the quality of the stone and mortar used, will cause it to vary greatly. For ordinary cut stone, we may with safety assume the angle of friction at 30° with the horizon : when laid in thin tempered mortar it is increased to 34° or 36°, and with very porous stone, such as freestone, laid in full mortar it will reach almost 45°.

It is to be observed, however, that this is not strictly true unless the arch is of sufficient thickness at bottom to prevent all tendency to upset inwards. A thickness of $\frac{1}{10}$ the radius of curvature is usually adopted as sufficient for this purpose.

Adopting 30° as the angle of repose for cut stone, the number of voussoirs which load the centre will depend on the curve given to the intrados. If we take, for instance, a full centre, an oval and a flat segmental arch, and give to each the same number of voussoirs, it is evident that the number of stones which do *not* press on the laggings will be greatest in the full centre, less in the oval, and least of all in the flat segmental arch, because in this latter case the stone whose lower joint makes an angle of 30° with the horizon will be found nearer the springing line. We should expect, therefore, the number and weight of the stones being the same, that the segmental arch could give the greatest load to the centres, and the full centre arch the least; and this is strictly the case.

In estimating the load upon the centres in any case, it is to be remembered that none of the stones bear upon the ribs with their entire weight, a part of this latter being consumed in overcoming friction. The determination of the amount of weight thus expended is a matter of some mathematical intricacy, and we are indebted for its solution to M. Couplet.* By his calculation he

* Mémoire de l'Academie Année, 1792.

found that the total weight of the voussoirs which *do* press on the laggings, is to the weight with which they actually load the frame, as an arc of 60° is to twice its sine less the same angle ; or, to express it algebraically, denote by *P* the total weight of the voussoirs which *rest* on the centring, and by *p*, the weight with which they *load* the centres, and we shall have the expression

$$P : p :: \text{arc } 60^\circ : 2 \sin 60^\circ - \text{arc } 60^\circ \qquad (1)$$

or

$$p = \frac{P\ (2 \sin 60^\circ - \text{arc } 60^\circ)}{\text{Arc } 60.} \qquad (2)$$

If, therefore, we suppose the radius of a circle to be divided into 10,000 equal parts, the circumference will contain 62,832, and the arc of 60° 10,472, and its sine is equal to $\frac{1}{2}\sqrt{3}$, 8660. Substituting these values in the above equation (1), we shall have

$$P : p :: 10472 :: 2 \times 8660 - 10472 \quad \text{or}$$
$$P : p :: 10472 : 6848$$

which gives us a ratio of 3 to 2 very

nearly. Whence we see that the voussoirs in a full centre arch which press upon the laggings will do so with but $\frac{2}{3}$ of their weight, and, taking the angle of repose on each side at 30°, only on $\frac{2}{3}$ of the surface of the centring. We may, therefore, without any sensible error take $\frac{4}{9}$ of the gross weight of the voussoirs of the arch to express the load on the centres.

With an arch which is not full centre the case is quite similar. We will take an oval of three centres fulfilling the conditions that each of the three arcs composing it shall be 60°. This oval being drawn, it is at once apparent that the arcs of 60° at each end of the oval do not differ materially from that of 30° in the full centre arch. We may, therefore, to facilitate calculation, safely assume that the stones forming these two arcs of 60° do not press on the centres, when the arch is all up except the keystone, and are held in place by the weight of the voussoirs above them. There remains then but the central arc of 60° to load the framing. But from equation (1) P : p as the arc of 60° is to twice its chord less the arc of 60° ; and since 60° is to its chord very nearly as 22 to 21, we may without sensible error express the relation of P to p by the ratio of 11 to 10. When we have found the gross weight of the voussoirs in this arc of 60° it follows that we must take $\frac{10}{11}$ of their weight to express the load on the framing.

The chord of an arc of 60° is equal to the radius, and the radius in this case being 10000, the *chord* will equal 10000, and the *arc* of 60°, 10472. Hence we have the relation 10000 : 10472 : : 21 : 22 *nearly.*

These values may also be obtained from the integral calculus, in which case no regard is taken of friction, and the formulæ are therefore a little uncertain. This uncertainty, however, is on the side of safety, for when we leave out of consideration the pressure expended in overcoming friction we are forced to give the ribs and laggings unnecessary strength.

Referring to Figure 1, we wish to find the load which the voussoirs between A B C D give to the centre's rib.

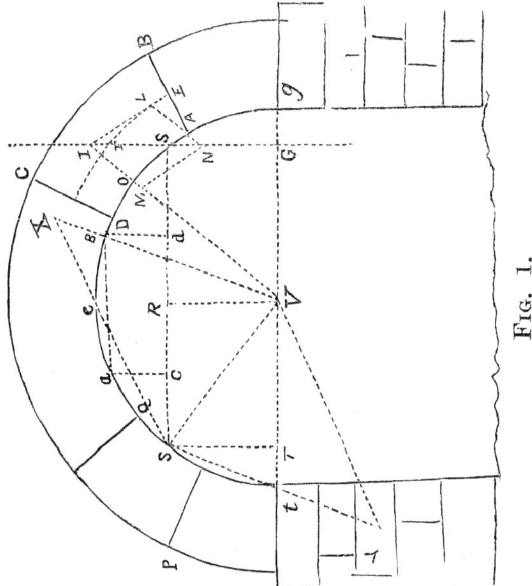

FIG. 1.

Let w equal the weight per lineal foot of intrados of the arch resting on the rib.

By x and y the horizontal and vertical co-ordinates respectively of the point A.

By x' and y' the co-ordinates of D.

By r the radius of the curve of intrados at A, and

By $\propto$ the angle it makes with the horizon.

By P the gross vertical pressure on the rib of A B C D.

By p the pressure per lineal foot of intrados at A.

Then we shall have

$$P = \int_{x'}^{x} p dx \quad . \quad . \quad . \quad (3)$$

From this we may compute the load on any vertical post at this point, or the vertical component of the load on the back pieces.

If the arch had been completed up to the keystone course, equation (3) would have been

$$P' = \int_{0}^{x''} p\, dx \quad . \quad . \quad . \quad (4)$$

P′ being the greatest vertical load on the *half* rib, and x'' the horizontal distance from the middle of the span to a point where p is zero.

The value of p is found from the equation

$$p = \text{W. Cos}\ \alpha - \frac{1}{r}\int_{y}^{y'} \text{W.}\,dy \quad . \quad (5)$$

which will evidently be greatest when

$$p = \text{W. Cos}\ \alpha \quad . \quad . \quad . \quad . \quad (6)$$

When the arc is a segment of a circle not greater than 120°, we shall have from the relation between the sides and angles of a triangle, the following values of the co-ordinates :

$$x = r \sin \alpha \qquad x' = r \sin \alpha' \quad x'' = r (\sin \alpha'')$$
$$y = r (1 - \cos \alpha) \qquad y' = r (1 - \cos \alpha')$$

Substituting these values in equation (5) we shall have

$$p = w (2 \cos \alpha - \cos \alpha'). \quad . \quad . \quad (7)$$

And from the expression $y = \text{r} (1 - \cos \alpha)$ we have

$$\text{Cos}\ \alpha = \frac{r - y}{r}\ ;$$

and from $y'=r\,(1-\cos\alpha')$

$$\text{Cos }\alpha'=\frac{r-y'}{r}$$

Substituting this in equation (7) we shall have

$$p=w\left(\frac{2\,r-2\,y}{r}-\frac{r-y'}{r}\right)$$

$$=w\,\frac{r-2\,y+y'}{r}\quad.\quad.\quad.\quad.\quad(8)$$

Equation (6) then becomes $p=w\cos\alpha$ $=w\frac{r-y}{r}$ the greatest value for a given point of the arch.

Substituting in equation (3) the value of p found in eq. (8), and, reducing, we obtain

$$\text{P}=wr[\alpha-\alpha'-\sin\alpha\,(\cos\alpha'-\cos\alpha)]\quad(9)$$

or

$$\text{P}=w\left(l-l'-\frac{x}{r}\,[y-y']\right)$$

in which l and l' represent the length in feet of the arcs from the crown E (Fig. 1) to the points A and D respectively.

Equation (4) then becomes

$$P' = wr\,(\alpha'' - \sin\alpha''\,[1 - \cos\alpha'']) \quad (10)$$

To find the gross weight of that portion of the arch which presses on the back pieces and laggings, it is necessary to know the number of the voussoirs, their volume and weight per cubic foot.

The weight of stone generally used in arches varies from 120 to 180 pounds per cubic foot. The following results were obtained from the examination of a number of specimens of American granite, sandstone and limestone, taken from the best known quarries in the country. Of seventy-two specimens of granite examined, the greatest weight per cubic foot was 182.5 lbs., the least 161.2, and the average 167.09 lbs. Of fifty-three specimens of sandstone examined, the greatest weight per cubic foot was 164.4 lbs, the least 127.5, and the average 140.9 lbs. Of thirty-eight specimens of limestone, the greatest weight per cubic foot was 173.8, the least 143.2, the average 162.9.

We may therefore without sensible

error assume the average weight of these three classes of stone as follows :

	Average weight per cubic foot.
Granite....	167.09 lbs.
Sandstone	140.9 lbs.
Limestone	162.9 lbs.
Brick (well burnt)...	93.0 lbs.

From the moment the angle of repose is passed and the first voussoir begins to press on the frames, the centring becomes subjected to a series of strains which increase rapidly up to the time the keystone is laid, and are produced by the yielding of the ribs under the weight of the stones. No matter how well seasoned and admirably proportioned the timbers may be, or how evenly the load may be distributed, the centre, pressed more and more severely on each side by the successive courses of voussoirs laid upon it, will bend in on the sides, and as a consequence bulge out at the crown, to be in turn followed by a bending in of the crown when the arch is all but completed. This movement of the ribs can be greatly checked and the severity of the resulting strains much lessened by loading the centres at the crown with the spare voussoirs and increasing the load as the arch progresses. In the case of a full centre arch of 90 feet, and composed of four hundred and eighty courses of voussoirs, the centring, when the fifteenth course of voussoirs on each side were laid in place, had risen three inches at the crown. When loaded with 325,000 lbs., it settled under it two inches; but when the twentieth course was completed the pressure was so great that it again rose one inch. When the arch was three-quarters completed it had again sunk one inch and three-quarters in consequence of the additional load and the compression of the wood, still leaving a rise of one quarter of an inch. This yield caused the joints at the twenty-second course to open a fraction of an inch, but closed when the keystones were driven. This distortion of the centring is always greatest for full centre arches, and proportionally less as the arch becomes nearer and nearer to the segmental.

DIRECTION OF THE STRAINS.

To find the direction and intensity of the strain at any point of the rib, we resort to the usual method of the "parallelogram of forces." Returning to Fig. 1, let it be required to find the direction of the strain caused by the voussoirs A B C D. Denote by F the centre of gravity of this part of the arch, and through it draw a vertical line G I of indefinite length, and cut it at I by a perpendicular from the point E at which the curve drawn through the centres of gravity of the voussoirs—supposed indefinitely small—cuts the line A B. Complete the parallelogram by drawing the line I M to the centre of arch, and N L parallel to it. The diagonal I N will then express the weight of the voussoirs A B C D, the side I L the pressure they exert upon the lower part of the arch, and the side I M the pressure upon the backpieces o the rib.

The strains, then, upon the centring take the direction of the radius of curvature of the intrados, and it now remains to consider the position which should be given to the beams which are to withstand the strains, their number and dimensions.

THE PRINCIPAL BEAMS AND THEIR POSITION.

As the sole object of the framing is to uphold the voussoirs and transmit the strains it receives as directly as possible to firm points of support, the beams must be so arranged as to do this with the least tendency to change the shape of the rib, by their bending or breaking. The condition will be best fulfilled by giving each beam a position such that it shall offer the greatest possible resistance, and this will be accomplished when the direction of the fibres of the beam and the direction of the strain are one and the same.

If, for instance, we support a horizontal beam at its two ends and load it in the middle it will offer its least resistance to the load. If now we raise one end so that the direction of the strain is oblique to the fibres of the beam, the resistance of the beam to bending will be found to have increased largely, and the resistance in this latter case, will be to that in the former case, as the cosine of the angle made by the direction of the strain and the fibres of the wood is to the sine of 90° or 1.

It should follow from this that, when the angle between the beam and the strain is zero, the resistance becomes infinite, and such would indeed be the case were it not for the compressibility of the wood and other physical causes which weakens its strength. It is sufficient, however, for us to know that when the strain is carried through the axis of the beam, it is then strongest, and that as the force becomes more and more oblique to the fibres its strength decreases.

Applying this fact to the framing of the ribs, it follows that the greatest stiffness and strength will be gained when the principal pieces are placed in the direction of the strains, or in the direction of the radii of curvature of the arch to be upheld. This deduction, unfortunately, is under certain restrictions placed upon it by the imperfections of the timber, and demands of economy and the circumstances of construction, which make its practical application quite limited.

To illustrate, we will once more return to Fig. 1. The direction and intensity of the strain on the backpieces resulting from the weight of the voussoirs ABCD, will then be represented, as we have just seen by the line V M, and that of the voussoirs P Q by the line V S. The beams, therefore, which are to support these stones, in order that they may offer the utmost resistance, must take the direction of the lines V S and V M, or radiate from the centre V like the spokes of a wheel. For small span arches, such an arrangement of beams undoubtedly answers all purposes of stiffness and economy, but for arches of larger span where timbers of thirty, fifty, or even a hundred feet in length would be required, it fails most signally; for while a beam of ten feet will offer great resistance to compression when loaded in the direction of the fibres, a beam of fifty feet will be almost sure to bend under the action of the strain, and hence require bracing. This system, therefore, cannot be successfully carried into practice in large span centres.

To overcome this difficulty we are forced to resolve the force represented by the line S V into two components, one vertical and represented by the line S T, and one horizontal represented in direction and intensity by S R. By a similar treatment of the force represented by V M, we shall obtain two other similar lines, all four of which will represent the direction of three beams, which can be made to take the direction of the two V S and V M, namely, a long horizontal beam spanning the arch and supported at each end by a vertical beam. This horizontal beam is the *tie* beam to which we have already alluded, and is generally placed at points about 45° up the arch. The voussoirs above this beam are then supported by another horizontal tie upheld by small vertical beams abutting on the lower tie. An excellent illustration of this system of framing is found in centres of London Bridge over the Thames, built in 1831 by Rennie.

There will frequently arise cases in which ribs framed in this manner either on account of the quantity of material they consume, or the difficulty of finding firm points of support between the abutments, cannot be used to advantage. It then becomes necessary to change the point of support T of the beam S T (Fig. 1) to a point *t* nearer the abutment, and for the sake of economy we may do away with the horizontal and vertical beams *t g*, *s* S, T S, *c a* and *a b*, supplying their place by two beams *t* S and S *e*. These two beams, therefore, will sustain the strain represented by the line S V, and the efforts they resist will be represented in direction and intensity by the sides S X and S Y of the parallelogram X Y constructed on S V as a diagonal.

In "cocket" centres, therefore, whatever the span of the arch, whether large or small, whatever the shape, whether full centre, oval or segmental, a great saving of material may be made, and abundance of strength may be secured, by placing the principal beams in the direction of the chords of the curve of the intrados.

The length that should be given to beams thus placed, the angle they should make with each other at their point of junction, the manner of supporting, and when necessary bracing them, are points we shall reserve for future consideration.

There are, therefore, three methods of arranging the principal pieces or struts of a centre frame.

1°. They may be placed in the direction of the radii of curvature of the arch, thus giving a figure of invariable form as the strain at any one point is received by the beam in the most favorable position, and transmitted through its axis directly to the fixed point of support.

2°. They may be placed in a vertical, or in vertical and horizontal directions.

3°. The curve of the arch may be divided into a number of arcs, and the beams placed in the direction of the chords of these arcs.

4°. To these three we may add a fourth, which embraces by far the largest number of centre frames, and is based on two or all of the preceding methods. In this class the beams are not arranged in accordance with any one system, but several; as, for instance, the second and third, in which case, as we shall see hereafter, several straining beams span the arch at different points, and are sustained by inclined struts; or if all three systems are used, we may use the straining beam and inclined struts, and strengthen them by bridle pieces in the direction of the radii.

It would, indeed, be quite a hopeless task to attempt to lay down, in more than a general way, the principles which ought to rule in making a selection of one of these methods to the exclusion of the remaining three. In every case the choice must be determined largely by the circumstances of the case, the points of support, the shape and span of the frame, and the strength required. If the centre is to be "cocket," the arch heavy, the span large, and considerable headway required beneath the frame, the third or fourth arrangement will undoubtedly afford the best results whatever may be the shape of the arch. If the arch is light, the span moderate, and little or no headway is wanted, then the second or first will generally be most convenient.

Theoretically, the first method will in all cases afford the greatest amount of strength and stability with the least amount of material, since the beams are then capable of resisting the most severe strains. Nor can there be any doubt that, within moderate limits, this result actually is attained in practice, and that of two ribs constructed with the same number of beams, of the same quality of wood and similar dimensions, in one of which the pieces are placed radially, and in the other vertically or inclined, the rib arranged on the former plan will be decidedly the stronger of the two. But, unfortunately, the impossibility of always obtaining firm points of support at the centre of curvature, the difficulty of finding sound, well seasoned timber of such length as would be required in arches of large span, and the relation which exists between the length and strength of beams under longitudinal compression—the strength varying inversely as the square of the length—restricts its application to centre frames of very small span and rise. In semi-circular arches of twelve, fifteen or even twenty feet span, when a horizontal beam can be used at the springing line this arrangement can be used with great success. The frame then consists of the tie beam and two, or if great strength is required, three radial struts which support the backpieces and abut against the horizontal beam at the centre of curvature. These struts, when two are used, should be inclined on the right and left at a little less than 45° to the horizon, so as to meet the backpieces at the point where the voussoirs first begin to press on the rib. A vertical strut is in such an arrangement of little or no use, as no strain of any consequence can possibly reach it ; the voussoirs almost ceasing to press on the frame when the keystone is driven down. As these supports are struts and not bridle pieces clamping the backpieces and tie beam between them, the joints, especially in the larger and heavier arches, must be secured by pieces of iron placed across them and bolted to the backpieces and struts, to prevent the joints opening in consequence of the bulging at the crown as course after course of stone is laid on the frame.

In frames for flat segmental arches of a span as great as sixty or seventy feet and rise of about one-fifth the span, as also for ovals of several centres, this radial arrangement may be slightly modi-

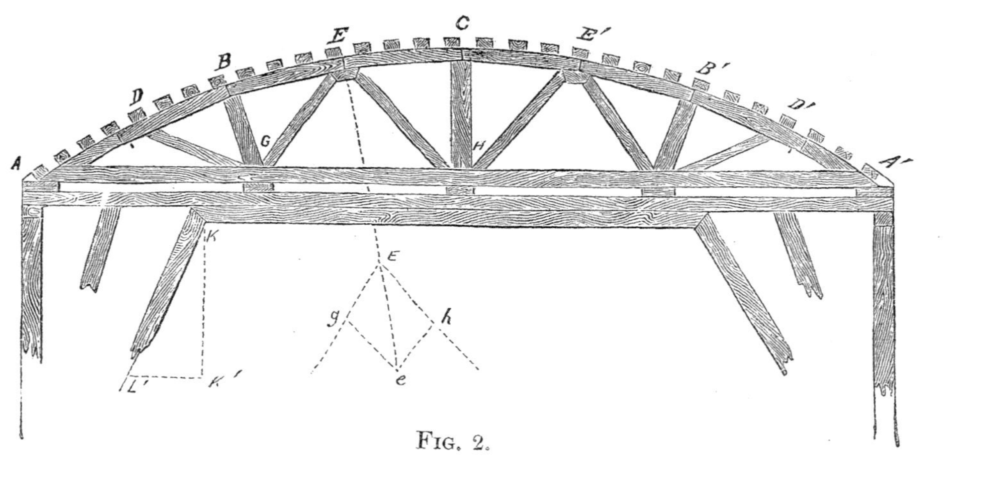

Fig. 2.

fied and a frame produced (Figure 2), which shall meet all the requirements of strength, lightness and economy. The rib in this case again consists of a horizontal tie beam spanning the arch a little above the springing line, generally at the first voussoir that presses on the backpieces, and struts placed in the direction of the radii of curvature and from eight to ten feet apart depending on the weight of the arch. These struts, as it would be impossible to have them actually meet at the centre of curvature, which, for an arch of seventy feet span and fifteen feet rise, would be about forty-five feet from the circumference, go no further than the tie beam and are fastened to it and the backpieces by the iron bands shown in the figure.

When great stiffness is required in the rib, additional braces may be added, as shown in Fig. 2, dividing the rib into a number of triangles. The strains received will then be transmitted through the axes of the beams, and as all unnecessary transversal strains will be avoided, the resistance offered by the braces will be the greatest possible. In all centre ribs, the *normal* pressure being in the direction of the radii of curvature, the laggings, backpieces and tie beam, when used, will of necessity be subjected to transversal strain.

Before, however, we proceed to consider the strains to which the beams in centre frames are subjected, and the dimensions we must give them in order that they may withstand the pressure put upon them, we would offer the following practical rule for estimating the pressure of any arch stone in any part of the arch, upon the centre rib, or the pressure upon the rib at any stage of the construction of the arch, as also the pressure when the arch is completed up to the key stone.

It has been well established by the experiments of Rondelet, that a stone placed upon any inclined plane does not begin to slide on that plane until it has reached an angle of inclination to the horizon equal to 30°. It is obvious, therefore, that if the arch stones were placed upon one another they would not begin to press on the centre rib till the plane of the lower joint of one of them reached an angle of 30° with the horizon. It has been found, moreover, that the mortar increases this angle, for hard stone to 34° or 36°, and for soft, porous stone (in semi-circular arches) to 42°. We may, then, consider the pressure to commence in general at the joint which makes an angle of 32° with the horizon. If we suppose the radius to represent the pressure the tangent will then represent the friction, and making the radius unity the friction will be 0.625. The next stone will press a little more, the third still more, and the pressure will thus continue to grow larger and larger with each succeeding course. The relation between the *weight* of an arch stone and its pressure upon the rib in the direction perpendicular to the curve is given by equation :

$$Q = W (\cos a - f \sin a) \quad . \quad . \quad (11)$$

in which Q is the pressure, W the weight

of the arch stone, f the friction = 0.625, and a the angle the lower joint makes with the vertical. The following table calculated from eq. 11, gives the value of Q for every 2° of curve from the angle of repose = 32° up to 60°:

When the angle which the joint makes with the *horizon* is

	34°.....	then	Q=.04 W
When	36°.....	"	Q=.08 W
"	38°.....	"	Q=.12 W
"	40°.....	"	Q=.17 W
"	42°.....	"	Q=.21 W
"	44°.....	"	Q=.25 W
"	46°.....	"	Q=.29 W
"	48°.....	"	Q=.33 W
"	50°.....	"	Q=.37 W
"	52°.....	"	Q=.40 W
"	54°.....	"	Q=.44 W
"	56°.....	"	Q=.48 W
"	58°.....	"	Q=.52 W
"	60°.....	"	Q=.54 W

To take an example: What is the pressure on a backpiece of 20° in length from the angle of repose, the ribs of the frame being placed 5 ft. from centre to centre, and the arch stones 3 ft. in depth and weighing 160 lbs. per cubic foot. We take from the above table the sum of the decimals from $32^\circ - 52^\circ = 2.26$, and multiply this by the weight upon 2° and the product will equal the pressure. The volume of the stones which cover $2^\circ = 5 \times 3 \times 2^\circ$.

The number of feet contained in 2° is found from the expression $2 \times .01745329 \times r'$, in which r' is equal to the radius of the arch *plus one half* the *depth* of the *arch stone.* If we take the radius $= 25$ ft., then the depth of the stones being 3 ft, $r' = 26.5$ and number of feet in 2° equals .88 ft., whence the volume of the stones which press on the 2° equals $5 \times 3 \times .88 = 13.4$ cubic feet, and the quantity $W = 2144$ lbs. and Q, or the pressure on the backpiece equals 4845 lbs.

If we denote by a the angle included between the upper and lower joints of an arch stone, and suppose every stone in the arch to have the same weight and equal angle a, then the pressure of any number n of such stone upon the rib will be given from the expression

$$Q = \frac{W + \sin \frac{n+1}{2} a \times (\cos \tfrac{1}{2} na - f \sin \tfrac{1}{2} na)}{\sin \tfrac{1}{2} a} \qquad (12)$$

which gives the total pressure on *one half* of the rib.

This equation is found as follows: The pressure perpendicular to the soffit is W ($\sin a - f \cos a$), or W ($\cos a - f \sin a$), according as the angle a is measured from the *horizon* or from the *vertical* drawn through the crown. If now we denote by a the angle included between the joints of *one* stone, and suppose each stone alike in size and weight, the pressure of any number n of such stones will evidently be found by getting the sum of the *sines* and *cosines* of $n\,a$, or expressed in formula,

$$P = W \text{ (sum of cosines of } n\,a - f \times \text{sum of sines of } n\,a) \quad \ldots\ldots \text{ Eq. A.}$$

By trigonometry we obtain two expressions for the sum of the sines and cosines of a number of angles in arithmetical progression, viz. :

$$\text{Sin } A+\sin (A+B)+\sin (A+n\,B) = \frac{\text{Cos } (A-\frac{1}{2}B)-\cos (A+n+\frac{1}{2}B)}{2 \sin \frac{1}{2} B}$$

$$= \frac{\text{Sin } (A+\frac{1}{2}\,n\,B)\times\sin \frac{1}{2}\,(n+1)\,B}{\sin \frac{1}{2}\,B.}$$

Also

$$\text{Cos } A+\cos (A+B)+\cos (A+n\,B) = \frac{-\cos (A+\frac{1}{2}\,n\,B)\times\sin \frac{1}{2}\,(n+1)\,B}{\sin \frac{1}{2}\,B.}$$

Applying these two equations to the above case, we shall have from eq. A,

$P=W$ Eq. B.

$$\frac{\cos \frac{n}{2}a\times\sin \frac{n+1}{2}a-f(\sin \frac{n}{2}\,a\times\sin \frac{n+1}{2}a)}{\sin \frac{1}{2}\,a.}$$

Or taking out the common factors W and $\dfrac{\sin \frac{n+1}{2}a}{\sin \frac{1}{2}a}$ we shall have equation B in the form. Eq. (12)

$$P=\frac{W\times\sin \frac{n+1}{2}a}{\sin \frac{1}{2}\,a}\times\left(\cos \frac{n}{2}\,a-f\sin \frac{n}{2}a\right)$$

The value of Q may also be obtained from eq. 11 by considering that when the depth of the arch stone is nearly double its thickness ; its *weight* rests on the rib at the angle of 60°. Equation 12 is, however, the best, and may be readily solved by logarithms.

For example : let the arch be semicircular and $a=2°$, then $n\,a=29°$ and $f=.625$. Put equation 12 in the form

$$P=W\left\{\frac{\cos\frac{1}{2}n\,a\times\sin\frac{n+1}{2}a}{\sin\frac{1}{2}a\times R}-\frac{f\sin\frac{1}{2}n\,a\times\sin\frac{n+1}{2}a}{\sin\frac{1}{2}a\times R}\right\}$$

$$\log\cos n\,a=\log\cos 29°=9.941819$$

$$\log\sin\frac{n+1}{2}a=\log\sin 30°=9.698970$$

$$19.639789$$

$$\log\sin\tfrac{1}{2}a=\log\sin 1°=8.241855$$

$$R=10.000000$$

$$18.241855$$

Difference= 1.397934
= log 24.68

$\log f$ = log .625= – 1.795880
$\log \sin \frac{1}{2}\, n\, a = \log \sin 29^\circ = 9.685571$
$\log \sin \frac{n+1}{2} a = \log \sin 30^\circ = 9.698970$

19.180421

$\log \sin \frac{1}{2}\, a = \log \sin 1^\circ = 8.241855$
R =10

18.241855
Difference= 0.938669 =
log 8.55

Hence the weight on the half rib is 24.68—8.55=16.13 W.

In a frame constructed, as that shown in Fig. 2, the determination of the strains is a matter of great simplicity, and may be had either from arithmetical calculation or by constructing the parallelogram of forces. The strain on any radial strut as B G would be found by calculating from eq. 11 the pressure on D E, taking half of it and supposing it to act at B in the direction B G. The strain on any inclined strut, as E G or E H, may be found by estimating from eq. 11, the strain on B H taking one half of it, and supposing it to act at E in the direction of the radius at that point, and denote by δ and δ' the angles these pieces make with the direction of the force. Then, if these angles are unequal

$$S=\frac{P \sin \delta'}{\sin (\delta+\delta')} \quad \text{and } S'=\frac{P \sin \delta}{\sin (\delta+\delta')} \quad (13)$$

And if the two beams make equal angles with the direction of the force, then the strain in the direction of each is the same and expressed by

$$S=\frac{P}{2 \cos \delta} \quad . \quad . \quad . \quad . \quad . \quad (14)$$

Of all methods of calculating the strain on the different beams, by far the simplest, is to actually construct the diagram of forces to a given scale and find the pressure by measurement. In above case, for example, draw E e parallel to the direction of the force to any convenient scale, say $\frac{1}{10}$ inch equal 1,000 lbs., which, supposing the pressure at E=10,000 lbs. will make E *e*=one inch. From E draw E *g* parallel to E G; also E *h* parallel to E H, and *e g* to E *h* and *e h* to E *g*. Then E *g* being measured will give the pressure on the beam E G to which it is drawn parallel.

When we have once ascertained the strain which any beam in a frame will have to undergo and resist, the next step is to determine the dimensions, or rather the area of cross section, the beam must have to withstand this pressure without injury. Whatever may be the length of the beam, this section may be obtained from the following formulæ: If the strain is one of compression in the direction of the length, then

$$A=\frac{F}{K}$$

in which A is the section required in square inches, F the crushing force to which the beam is subjected, and K the *resistance* to *crushing*. When the strain is a transverse or breaking strain, then

$$A=\frac{F}{K'}$$

in which K′ is the *modulus* of *rupture* of the beam.

In place of K and K′, however, which are the ultimate resistance to crushing or rupture, we must use $\frac{K}{n}$ and $\frac{K'}{n}$, in which n is the factor of safety, usually taken as 10 for wood. The values of K and K′ are variously stated by different writers on the strength of materials. Those given below for the woods mostly used in centre frames are from Rankine:

Wood.	Value of K in lbs	Value of K′ in lbs.
Ash	9,000	12,000
Pine, yellow	5,400	9,900
Pine, red	6,200	7,100
Oak, English	10,000	10,000—13,000
Oak, American	6,000	10,600

If it is not always possible to obtain these values of K and K′, a very safe method, and one easily remembered, is to find from the diagram of forces the strain on a beam in lbs., and divide this by 1,000; the result will be the cross section of the beam in inches. Thus, if a timber is loaded with 36,000 lbs., $\frac{36,000}{1,000}=36$ in., and the beam should be 6 in. × 6 in.

Example.

Required the proper dimension of the scantling of a centre rib of a segmental arch of 60 feet span and 9 feet rise ; the arch stones to consist of old quarry granite, weighing 165 pounds per cubic foot, and three feet in depth ; the rib to be of the pattern shown in Fig. 2. The frames to be placed 5 ft. from centre to centre.

The first step is to find the weight of the arch stone for 1° of the curve. The span is 60 ft., the radius is 50 ft., and the arch stones being 3 ft. thick the radius of the arch passing through their centre is 51.5 ft. The length of 1° is, therefore, .01745329 × 51.5 = .89 ft. Then 5 × 3 × .89 = 13.3

cubic ft., the solid contents of 1° of the arch ring, and this multiplied by 165 gives the weight of $1° = 13.3 \times 165 = 2195$ pounds. Now the arch being a very flat segmental, it is evident that *all* the arch stones will press upon the rib. If then we calculate the weight of the stones between E E′, and suppose them to act with one half their entire weight at C in the direction C H, it is evident that this will be the greatest pressure that C H will be required to support. The arc E E′=20°, and the weight for 1° being 2195 lbs., the pressure at C is 21950 lbs., and the beam C H should be

$\frac{21950}{1000} = 21.9$ inches or $4\frac{1}{2} \times 5$ in. To find the dimensions of E G and E H take eq. 12. Then $a = 1°$, $n = 20°$, $f = .625$, $W = 2195$.

$$Q = \frac{2195 + 182236}{.008727} \times (984808 - .625 \times 173648) = 18301 \text{ lbs.}$$

Take this and lay it off to any convenient scale on the line E e, and from E draw E*g* parallel to E G, and E*h* to E H and as before *e g* and *e h*. Then measuring E *h* by the same scale it will be found to equal 10250 lbs. ; the beam E H then must be 3½ in. by 3 in. In the same manner the pressure on B G is found to be 18301 lbs., and the beam must be 4½ in.×4 in. To find the strain on the inclined strut, estimate from eq. 12 the weight of the arch stones between A and C, add to this half the weight of the rib and let the gross weight act vertically at the point K, and lay it off to any scale on the vertical line K K′, and draw K′ L′ parallel to the horizontal tie beam. The line K L′ being measured will give the strain on the beam K L′.

Frames arranged on the second method, with the principal pieces all vertical, afford centres of great simplicity of structure and of almost as much strength as one with radial struts—supposing, of course, that the number and dimensions of the struts are the same in each case—and of much greater strength than one constructed with inclined beams, since the nearer the angle the direction of the strain makes with the fibres of the wood approaches a right angle the less becomes the resistance of the beam. In segmental and oval arches of large span, the difference in the strength of ribs arranged on the vertical and radial plan is comparatively insignificant, as the radius being very large, the vertical beams, especially near the crown where the strain is severe and most strength is required, do not depart much from the direction of the radius.

The objection to this vertical bracing of the frame is that it requires the use of a horizontal tie beam, unless the rib is constructed as a girder resting upon framed abutments of its own. If the former arrangement is used, the struts should be placed from five to eight feet apart, depending on the strength required, and mortised to the tie beam and backpiece. When the beams are of such length that there is danger of their bulging or curving under the load laid on them, they may be strengthened by diagonal braces or horizontal wales. Of the two, the diagonal braces are to be preferred as they not only give stiffness to the posts, but sustain a portion of the load on the backpieces in case any of the piles under the horizontal tie beam should give way. Figure 3 represents the rib of a full centre arch of 75 ft. span arranged with the principal pieces placed vertically and strengthened with a horizontal waling piece made double, and braces abutting under the backpieces. The strains on the different beams composing such a frame, and their necessary dimensions may be computed with ease by the method just explained. It should, however, be remembered that beams which are to be notched must have their dimensions increased beyond those given by calculation, in as much as notching will, even when not very deep, cut down the strength of a beam from one third to one half. In computing the strains on the braces a, a, we may consider the pressure at their abutting point to be the sum of the pressures on the vertical

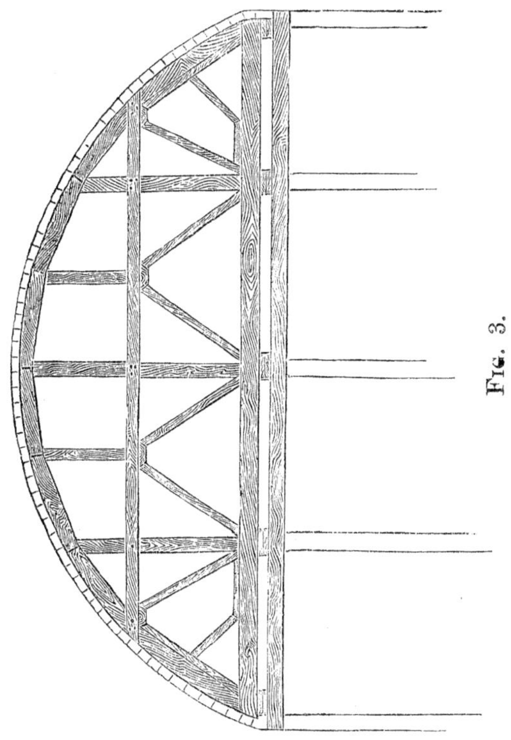

Fig. 3.

and two inclined braces which meet there, and make no allowance for the resistance of the horizontal beam.

The third and fourth systems of arranging the principal pieces, afford an almost unlimited number of designs for centre ribs, which are especially worthy of notice, in that they are applicable to every possible shape and span that can be given to stone arches, and may be constructed with or without intermediate points of support, according as circumstances will admit. The principles which control such arrangements are few and simple. The beams should as far as possible abut end to end: they should intersect each other as little as may be since every joint causes some degree of settlement, and halving destroys fully half the strength of the beams halved. When the framing is composed of a number of beams crossing each other, pieces tending towards the centre should be notched upon and bolted to the framing in pairs : ties should also be continued across the frame at points where many timbers meet. Particular attention must, furthermore, be given to the manner of connecting the beams so that there shall be no tendency to rise at the crown under the action of the varying load, Figure 4 affords an illustration of a very simple method of arranging the timbers for arches of small span. The inclined struts abut against horizontal straining beams placed at different points on the soffit, and to add greater strength to the framing, and to prevent the horizontal beam from sagging, bridle pieces are placed in the direction of the radii of curvature. The chief difficulty with such arrangement as this is, that as they require beams of great length they can be used to advantage only in small span arches.

The centre frames for the Waterloo Bridge over the Thames were constructed on this principle, but in this case no horizontal beams were used. Under the backpieces were placed blocks each supported by two inclined struts which made equal angles with the radius drawn

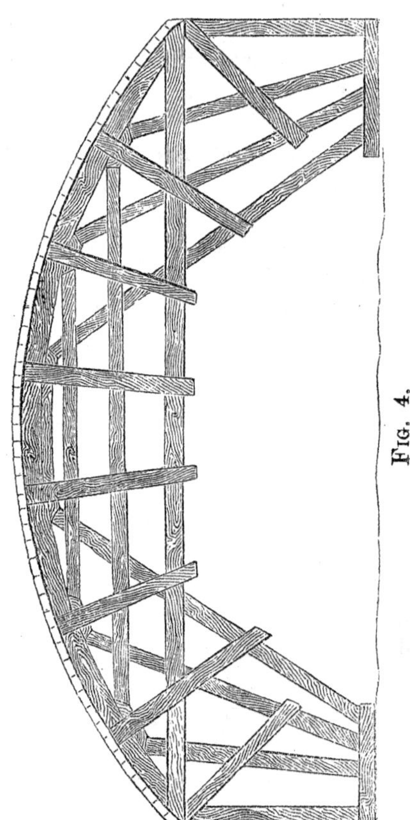

Fig. 4.

through the centre of the block. In a small span arch, these struts would have rested on framed supports placed at the opposite abutments of the arch ; but in the Waterloo Bridge, to avoid the inconveniences resulting from crossing the struts, and of building beams where struts of sufficient length could not be obtained from single beams, the ends of several struts were received into cast-iron sockets placed at their point of crossing and suspended by bridle pieces.

Figure 5 is a good design for a cocket centre of large span. Here the C F′, H F and D *d*, are placed in the direction of the radii of curvature and made double ; the remaining braces are single. In determining the proper dimensions for the scantling of such a frame, we may take $\frac{2}{3}$ of the total pressure on the arc H H′, and suppose it to act at C in the direction C F′, which will evidently be the greatest load this timber will have to sustain. The strain upon the E D F will then be equal to $\frac{1}{2}$ the load on B H, and that on H F as $\frac{1}{2}$ D C. That on the

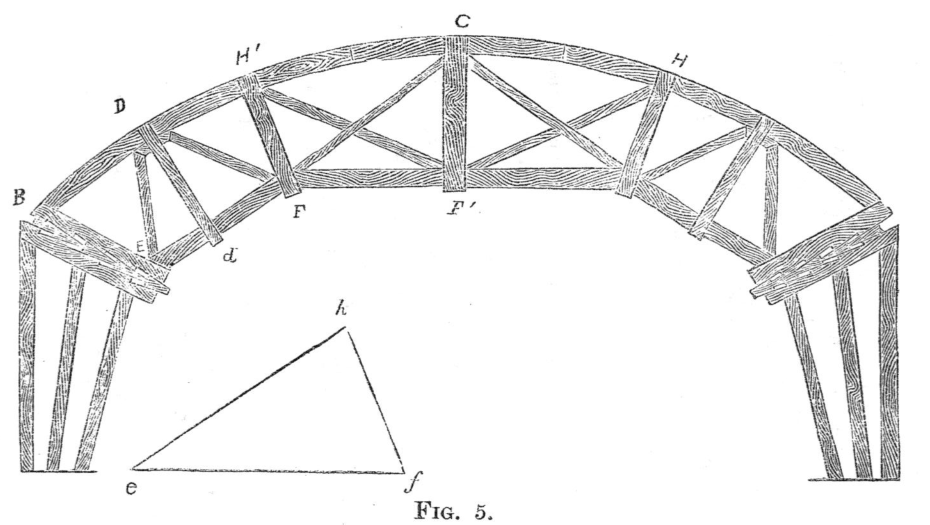

FIG. 5.

beams E F and F F′ is to be found from the diagram of forces, Fig. 5. Here hf which is in the direction of H F produced, represents the pressure on this beam; Eh is drawn parallel to E F, and ef parallel to F F′, which being measured give the strain on E F and F F′ respectively. If it is desirable to obtain the dimensions of the beams with great accuracy we may use the following formulæ: If we assume the relation between the breadth and depth to be .6 to 1 (which is an excellent proportion), then for an inclined beam whose angle of inclination to the horizon is β.

$$d = \sqrt{L\sqrt{\frac{W. \times \cos\beta \times a}{0.6}}} \quad . \quad . \quad (15)$$

And for a horizontal beam

$$d = \sqrt{L\sqrt{\frac{W. a}{0.6}}} \quad . \quad . \quad . \quad . \quad . \quad . \quad . \quad (16)$$

In which a is to be found from the expression $\frac{40 \times b \times d^3 \times \delta}{L^3 W} = a$, in which δ is the deflection of a beam whose breadth is b, depth is d, length L, and load W. For pine this quantity a is from .0112 to .0105, and for the best oak .00934. Eq. 15 or 16 will give the depth in inches. If it so happens that the value of a, in the above equation, cannot be obtained either by actual experiment or from tables, we may make the square of one side equal to twice the square of the other, which will give a ratio of 7 to 5 very nearly, and use the equation

$$d = 0.0046108 \times \sqrt[3]{\frac{w.\,l^2}{\cos b}}$$

Where w is the load, l the length, and b the angle the beam makes with the vertical, and d the dimension of the *smaller* side, equal $\frac{5}{7}$ of the larger. In centre frames, however, such a degree of exactness is rather unnecessary, since, by allowing 1,000 lbs. to the square inch we can obtain the cross section from the load with all the accuracy desirable in practice.

The transversal strain on any one back-piece or segment of the rib under the laggings may be obtained from the expression

$$S=P \sec \phi \quad . \quad . \quad . \quad (17)$$

ϕ being the angle the backpiece makes with the horizon, and P the vertical component of the pressure on the same piece found by any of the methods already explained, or from

$$P=W\left(L-\frac{x}{r}h\right) \quad . \quad . \quad . \quad (18)$$

W being the pressure on each lineal foot of the segment, L its length; r the radius of curvature at the point in question, x the distance of the lower end of the backpiece from the vertical through the crown of the arch and the centre of curvature, and h the distance between the two ends of the segment measured vertically.

The strain upon any one of the laggings will depend, independent of the weight of the arch stones, on the distance of the ribs from centre to centre, the place the lagging occupies in the arch and the manner in which the laggings are attached to the backpieces of the frame. As regards the latter point, there are two ways of making them fast to the rib. They may be placed directly on the backpiece and nailed to it, or they may be mounted on folding wedges placed between each bolster or lagging and the rib, which latter arrangement will be considered in detail when we come to speak of the striking plate. The bolsters, moreover, may be placed on the rib in such wise that they touch each other, or may be separated by a space equal to their own breadth. The former method is most usually resorted to in the construction of brick arches, and is illustrated in Fig. 4 ; the latter is used in building stone arches, and is illustrated in Fig. 2. By separating the laggings in this wise a considerable saving of timber is effected, while the air is also given freer access to the joints of the arch and the mortar much sooner dried. When these pieces are separated, it is evident that the cross section of each must be slightly greater than when they are placed touching each other, and that the section of the laggings placed near the crown should be larger than those near the angle of repose. This latter point is not worth considering in practice unless the arch stones are very heavy, for in arches of the ordinary span and weight the saving thus effected in the timber is hardly worth the labor of calculation. In determining the proper dimensions of the laggings, it is sometimes customary to insure against any deflection, by supposing the entire load on each lagging to act at its middle point and calculate for a beam strained in this manner.

BRACING.

It is to be observed in connection with the matter of bracing, that the frames should be arranged in such wise that no piece suffers any strain other than *compression* or *extension* in the direction of its length. As it is, however, by no means an easy matter to make the distinction, we shall give the following rule to which there is no exception :

Suppose we have two beams abutting against each other at their upper end, and loaded at their point of intersection with a weight. Take notice of the *direction* in which this *straining force* acts, and *from* the point *at* which it acts draw in this direction a line representing by its length the intensity of the strain. From the remote end of this line draw lines parallel to the two pieces on which the strain is exerted. The line drawn parallel to one must of necessity cut the other or its direction produced. If it cut the *beam itself* the piece is *compressed*, and acts as a *strut.* If, on the other hand, it cuts *the direction of the beam produced*, the piece is *stretched* and acts as a tie. We may then lay it down as a general rule in framing, that if the piece *from which* the strain comes lies *within* the angle formed by the pieces strained, the *strains these* sustain are of the *opposite* kind to that of the *straining point;* if that is *pulling*, they are pushing; if that is *compressed* they are *stretched.* Again, if the piece from which the strain comes lies *within the angle formed by the direction of the two produced*, *all* will have the same kind of strain ; and, finally, if within the angle formed by the *direction of one produced* and the *other piece itself*, the strain will be of the *same kind* as that of the *most remote* of the two beams strained, and of the *opposite* kind to that of the *nearest.*

The object of all bracing, then, being to convert all transversal strains into others which act in the direction of the length of the beams, the frame must be divided into a number of triangles ; for as the triangle, or some modification of it, is the only geometrical figure which possesses the property of preserving its figure unaltered so long as the length of its sides remain constant, it is the figure best suited for structures in which rigidity is essential for stability. But, again, some forms of triangles are much to be preferred to others ; the strength of the pieces forming the triangle depending very much on the angle they make with each other. *Oblique* angles are to be avoided. *Acute* angles when not accompanied by oblique are not so injurious, because the strain can, in such pieces, never exceed the straining force ; but in an *oblique* angle it can surpass it to any degree.

In all forms of bracing, too much attention cannot be given to the joints. Where the beams stand square with each other, and the strains are also square with the beams and in the plane of the frame, the common mortise and tenon is the most perfect joint, a pin usually put through both so as to draw the tenon sight into the mortise, and so cause the thoulder to butt very snugly. Round pins are much better than square ones, as they are not liable to split the bit. Where the beams are very oblique, it is difficult to give the foot of the abutting one such a hold as to bring many of its fibres into actual contact with the beam butted on. It would, in such case, seem proper to give it a deep hold with a long tenon. Nothing, however, can be more injurious, for experience has fully proved that they are very liable to break up the wood above them and push their way along the beam. For instance, suppose the head of an inclined strut abutting on a horizontal beam to descend a little; the angle with this latter beam is diminished, by the strut revolving round the stress in the tie beam. By this motion the bed of the strut becomes a powerful fulcrum to a very long lever ; the tenon is the other arm and very short. It therefore forces up the wood above it and slides along the horizontal beam. This may be prevented by making the tenon shorter,and giving to its toe a shape which will make it butt firmly in the direction of the thrust, on the solid bottom of the mortise. When the beam is a tie the joint must depend for its strength on the pins or bolts, and the iron straps placed across it.

STRIKING THE CENTRES.

Undoubtedly the most dangerous operation connected with the use of bridge centres is the process of striking them.

No matter with how much care the arch may have been constructed, the drying and squeezing of the mortar will cause it to settle in some degree when the centres are removed, and this degree of settlement seems to be very largely affected by the time the centres are allowed to stand. By some it has been urged that the centring should never be removed until the mortar in the joints of the last course has had ample time to harden; others going to the other extreme have advocated striking the ribs as soon as the arch is keyed, claiming, not without some reason, that the settlement of a *well built* arch will never be so great as to become dangerous even though the supporting frames be removed when the mortar is green. But possibly the best practice lies not far from either of these extremes. It has, indeed, time and again, been amply demonstrated that to leave the centring standing till the mortar has hardened, and *then* take away all support, the mortar having become unyielding, is to cause the courses to open along their joints. To strike the centre, on the other hand, when the arch is green will, seven cases out of ten, be followed by the fall of the bridge ; but by easing the centring as soon as the arch is keyed in, and continuing this gradual easing till the framing is quite free from the arch, the latter has time to settle slowly as the mortar hardens, and the settlement will be found to be very small.

It becomes necessary, therefore, to provide some arrangement by which the framing may be slowly lowered from the soffit of the arch, an operation accomplished in a variety of ways ; by folding or double wedges, by striking plates, by bearing irons and screws, by cutting off the ends of the principal supports, and, finally, by plate iron cylinders filled with sand. The folding wedges are, perhaps, most commonly met with in practice, and are finely suited for arches of small span, as a sill stretching from abutment to abutment may then be used to rest them on. They consist of two hardwood wedges, about 15 in. long, right angled along one edge, and placed one upon the other in such wise that the thick end of one shall be over the thin end of the other, thus making their surface of contact an inclined plane. These wedges are placed under the tie beam of the rib and on the sill, as is illustrated in Fig. 2. It is evident that by driving the upper wedge up along the inclined surface of the lower, the rib which rests upon the upper one must rise, so that by placing a number of these folding wedges under each rib it may easily be keyed up to the desired level, and by driving the upper down the inclined surface of the lower, the rib may gradually be lowered. To keep the under wedge in place, it is usually made fast to the sill and the surface of contact of each wedge well greased with soft-soap and black lead. When the wedges are in place under the rib, it is a good practice to mark each wedge at the point where contact ceases, so that when the centres are being lowered we may be able to know whether they are lowered uniformly or not. For instance, let the lower wedges of three pair of folding wedges project two inches beyond the end of the upper ones, and mark with chalk on the *side* of each lower wedge the point where contact ceases; namely, two inches from its end. Now, if in striking the centres the upper wedges have *all* been driven back so that the *end* of each instead of being *at* the line is one inch *beyond it*, then the frame has been uniformly lowered; but if some are one inch and some $\frac{3}{4}$ inch from the line, the frame has not been lowered uniformly, and the difference must be corrected by driving *all* the wedges till they are one inch from the chalk line.

It is evident that such an arrangement of folding wedges can be of but little use unless the horizontal beam or sill on which they rest is rigidly supported from beneath, as any yielding of the sill would be followed by a separation of the wedges and rib. In constructing bridges of wide span over creeks or rivers on which there is no navigation to be interrupted, it is usual to make use of the folding wedges and support the sill by a row of piles driven into the river bed, and it then becomes especially necessary to watch the wedges lest by some settling of the piles and sill they have separated in the smallest degree from the tie beam of the rib.

In cocket centres the folding wedges are replaced by a *sriking plate* placed at each end of the rib, and sustained by strutting or raking pieces which abut either on off-sets at the foot of the pier or on sills placed on the ground. Each plate consists of three parts, a lower and upper plate and a compound wedge driven between them. The upper of these plates is of wood made fast to the base of the rib, and is cut into a series of offsets on its *under* surface (see Fig. 4). The lower plate is likewise of wood cut into offsets, but on its *upper* surface, and is firmly attached to the raking pieces which sustain it. The compound wedge consists of a beam cut into offsets both upon its upper and lower sides so as to fit those of the two plates, and when driven between them is held in place by keys driven behind its shoulders.

Previous to the time of Hartley, the rib was struck in one piece by the use either of wedges or striking plates. To him, however, we are indebted for an improved system of striking or easing the centres by supporting each lagging upon folding wedges. When this arrangement is used the rib is firmly attached to its supports, and the laggings rest upon wedges placed between them and the back pieces of the rib. A great advantage gained by this, is that the laggings may be removed course by course from under the arch, and replaced if the settlement prove to be too great at any one part of the soffit. Another method, at one time much in use among French engineers, is to cut off the ends of the chief supports of the rib piece by piece, an operation which cannot be accomplished with much regularity, nor without much danger.

The least objectionable way of striking centres, and one accomplished with great ease and regularity is by the use of sand, confined in cylinders. A number of plate iron cylinders one foot high and one foot in diameter are placed upon a stout platform sustained by timber framing. The lower end of each cylinder is stopped by a circular disc of wood of an inch thickness fitting tightly into the cylinder, and at about an inch above this wooden bottom three or four holes an inch each in diameter are drilled through the iron sides of the cylinder and stopped with corks or plugs of wood.

Into the cylinders thus prepared is poured clean dry sand to a height of 9 or 10 inches above the bottom, and on this sand in each cylinder rests a cylindrical wooden plunger, which fits so loosely as to work with ease, and forms one of the vertical supports of the rib. To prevent moisture getting at the sand, the joint between the plunger and cylinder is filled with cement. So long as the sand is dry it remains incompressible to any weight that may press on it, and the rib is thus kept invariably in its place. When the centre is to be lowered, the plugs are taken out of the cylinder, and as the sand runs out of each with *uniform* velocity the frame is *uniformly* lowered. This method is of especial value for centres of great weight.

The distance at which the frames or ribs of centres should be placed apart, measuring from the centre of one rib to that of the next, must be regulated solely by the weight of stone used for the arch, the distance varying inversely with the increase of weight. That is to say, if we assume some distance for stones of a given weight, say 6 feet for stones weighing 150 lbs. per cubic yard, and wish to find the proper distance apart of the ribs when the stones weigh but 120 lbs. per cubic yard, we have

$150 : 120 :: 5 : 4$. Then making 6 ft. the distance for 150 lb.,

$4 : 5 :: 6 : x \quad 4x=30 \quad x=7$ ft. 6 in.,

the proper distance for stones of 120 lbs. per cubic yard. The following table has been calculated in this manner :

Weight of Stone per Cubic Yard.	Distance apart of the Rib of Centring.
120 lbs	7 ft. 6 in.
125 lbs	7 ft. 3 in.
130 lbs	6 ft. 11 in.
135 lbs	6 ft. 8 in.
140 lbs	6 ft. 5 in.
145 lbs	6 ft. 2 in.
150 lbs	6 ft. 0 in.
155 lbs	5 ft. 10 in.
160 lbs	5 ft. 7 in.
165 lbs	5 ft. 5 in.
170 lbs	5 ft. 3 in.
175 lbs	5 ft. 1½ in.
180 lbs	5 ft. 0 in.
185 lbs	4 ft. 10¾ in.
190 lbs	4 ft. 8 in.
195 lbs	4 ft. 7 in.
200 lbs	4 ft. 1 in.

It now remains to consider briefly, the subject of centring as used in the construction of the arched roofs of tunnels. In work of this description, the span being always small, the arch light and the facilities for obtaining firm points of support for each rib as great as can be desired, all the hindrances, that so often make the framing of a stone bridge centre a matter of no small difficulty and foresight, are wanting, and the rib admits of a simplicity of arrangement at once favorable to economy of material and of space. It must, however, be remembered that although the span is small and the arch light, the strength of the rib of a tunnel centre must be much greater in proportion to the burden it has to carry than that of a bridge centre ; since the former has not only to resist the weight of the earth above it, but must also withstand the wear and tear of many destructful causes to which the latter is never exposed. In tunneling through a hill side, no matter how short the distance, more or less rock will invariably be met with, and more or less blasting must therefore be done, and the shock and flying splinters of rock which accompany each explosion do much mischief to the ribs by disturbing or injuring them. This cause acts strongly on all parts of the centre, but is especially severe with the leading ribs, which, as the brick work must always be kept well up to the heading, are directly exposed to the violence of each explosion.

A second cause of injury to the ribs, and one quite as damaging and unavoidable as the first, is the repeated taking down, carrying forward, and putting up of the ribs every time a length of arch is completed. In bridges, unless the structure is composed of a series of arches, the centring is never disturbed from the time it is first put up until it is finally *struck* on the completion of the works. In tunneling, however, to avoid the foolish expense of building centres from end to end of the tunnel, it is customary to construct but one length of twelve or fifteen feet of centring, and to move this forward whenever it becomes necessary to turn a new length of arch. Thus, for example, we will suppose that we are driving a tunnel through earth of a moderate degree of heaviness, and are, therefore, using centres consisting of two sets of laggings and five ribs, two made without and three with a horizontal tie beam. The object in making some of these ribs without the tie beam is that, by so doing, the centring may be brought close up to the heading without interfering with the raking props, which could not be done were the beams to be retained. These five ribs are arranged in practice so that one without the tie beam shall be placed at each end of the length of centring, and between these two are the three with beams. We will suppose this to be the arrangement of the ribs in the present case, and will number them, beginning with that nearest the heading, 1, 2, 3, 4, 5. While the arch is being turned upon this length the excavation for a new one has been made, the invert built, the side walls raised to springing line and all is ready to carry forward the centring. This operation, however, must be done with the utmost caution. If the ribs are taken from under the newly completed arch before the invert and side walls of the advanced length are built, the whole piece of arch with its side walls will be almost certain to separate from the length just behind, and move forward several inches in the direction the work is progressing. If, on the other hand, after the advanced side-walls are up, *all* the ribs are taken from under the arch, this latter will be quite certain to come down in ruins, since it has to uphold not only the weight of the earth resting immediately upon its bricks, but, in addition, *half* the *weight of the earth* which presses upon the crown bars of the newly excavated length, as one end of all these bars rests upon the arch near its end. Rib number 1, then, which is directly beneath the end of the crown bars, can not be removed with any degree of safety. It is also desirable that number 3 should be left in place to help support the laggings. Numbers 2, 4 and 5 are the only ribs left, and these are to be taken down and set up forward, taking care that 5, which has no tie beam, is placed nearest the heading; the order of arrangement then being 5, 4, 2, 1, 3.

Over the rib thus arranged a second set of laggings is laid, and on them the arch is turned. When this length is completed, and all preparation made to carry forward the centring, the ribs numbered 4, 1, 3 are taken down and set up forward in the order 1, 3, 4, 5, 2, and so on till the centring reaches the end of the tunnel, or meets that coming from the opposite end of the tunnel, supposing it to be worked both ways.

Now, it is precisely this continual taking down and setting up of the ribs, that produces so much injury to them, since, in order to pass them under the forward ribs and props which remain standing, it is necessary to take them in pieces. Each rib, therefore, must be framed in such wise that it may be repeatedly taken apart and put together again without injury to its strength or to the joints of the timbers removed and replaced. Figs. 5 and 6 afford an illustration of two centre ribs arranged to meet these requirements in the simplest manner possible. Fig. 5 is a drawing of a leading or segment rib, which it will be observed is constructed without a complete tie beam at the bottom so as to offer no obstruction to the raking props. It consists of two parts or segments, which, when the rib is placed, join at the crown of the arch and along the line *a b*, and are made fast to each other by two iron bars placed across the joint at the crown, one on each side of the back-pieces, and bolted through the back-pieces as shown at *c c*. An additional band is passed around the two vertical beams as shown at *d*. To prevent any slipping of these beams along the joint *a b*, the surface of each beam is notched, as shown at *e*, and a wedge driven through the notch. When the rib is to be taken down, the band at *c c* and that at *d* is removed, and the wedge at *e* driven out, and the rib thus separated into two segments may be carried through a comparatively small space. As this leading rib is subjected to the direct effects of the blasts, and to flying fragments of rocks, its joints must be strengthened by irons placed on each side of the rib, over the joint, and bolted through the timbers as shown in the figure.

This form of rib is finely adapted for tunnel centring, as it may be taken apart without removing a single beam, while its joint is so arranged that the pressure of the arch assists in no small degree to hold its parts in place. Indeed, the only valid reason why this form of rib should not be used in every part of a tunnel centre is the absence of the tie beam, which is certainly a great security against the spreading or contracting of the span. Were this tie beam supplied, and it may easily be supplied by an iron screw rod, this form of frame would probably, in addition to the convenience of taking apart and resetting, sustain any amount of pressure ever likely to occur either vertically or laterally, as also all ordinary wear and tear from use.

Fig. 6 represents one of the interme-

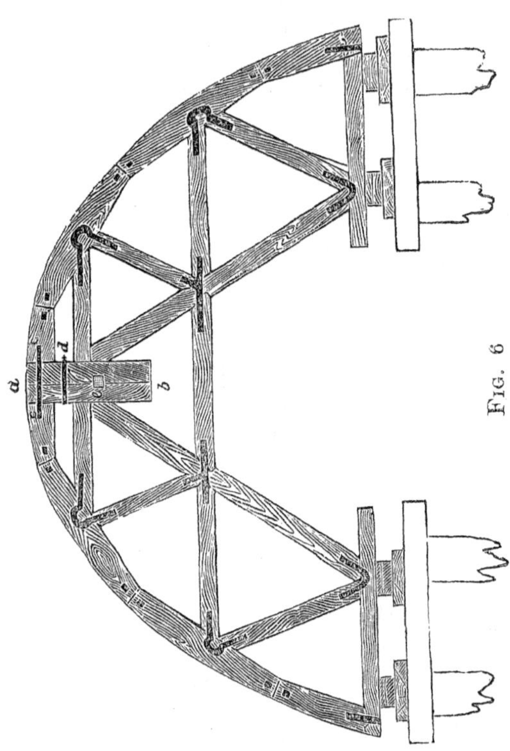

Fig. 6

diate ribs called scarf or queen post centres, which, as there are no props to be interfered with, are provided with horizontal tie beams. As these ribs are also to be taken apart each time they are shifted, the tie beam is composed of two beams joined by a scarf joint strenthened by a piece of timber placed above it, and bound to the tie by two bands of iron as shown in the figure. The horizontal beam joining the queen posts is also movable, and is held in place by the iron placed over its joints and bolted through. In joints thus protected, the holes through which the bolts pass are liable after a time to become so much enlarged, from the repeated driving in and out of the bolts, so as to injure the strength of the joint. This may readily be overcome by using a bolt with screw threads at each end in place of a bolt with a head and one nut, so that when once driven through the beam it need not be removed.

By a comparison of these two forms of ribs, it is evident that while the queen post centre possesses an advantage over the segment form in that it is not liable to lateral spread, it is at the same time inferior to the former in many important points. It cannot so well resist shocks or side blows, and being so taken to pieces every time it is moved is very liable to be injured especially at the scarf joint. An additional recommendation for centres constructed on the plans of Figs. 5 and 6, is the small amount of material used, which is quite as small as is consistent with the varying strains the ribs are exposed to, and is so cut that the timbers are almost as valuable when the tunneling is completed as they were when first purchased for the ribs.

The estimation of the dimensions proper to give each tie and brace of the rib is easily determined in so simple an arrangement, by any of the methods given for bridge centres. It is, however, to be remembered that, while the bridge centre has to sustain but the weight of the arch stones and bonding mortar, a load which can be calculated to a pound before one stone is laid, the centring of a tunnel has to resist the pressure not only of the brick roof, but also of the earth above, and that this latter pressure is wonderfully variable. The pressure of the brick work will of course vary when laid in cement and when laid in mortar. From the most careful experiments made to determine the weight of a cubic yard of brick work, we find that when the bricks are laid with cement the weight per cubic yard is 2,897 pounds, or in round numbers 2,900 lbs.; when laid in mortar beds the weight falls to 2,677, a difference of some 220 lbs. per cubic yard. It is true that the pressure of the earth does not act to any great extent on the centring, until the arch is turned and the crown bars drawn forward to form the roofing of the newly excavated length, but when this is done, and the three ribs removed to be set up in advance, the pressure on the two ribs remaining under the arch is quite severe. This load is especially variable with the leading or segment ribs, which it will be remembered are placed at the ends of the length of arch, and sustain one end of all the side and crown bars supporting the earth, and the movement which this earth is at any moment liable to take, cannot be foreseen. At times a whole length can be gotten out and the arch turned without any perceptible motion of the earth either at the sides or on top; at others, the earth will of a sudden begin to move and throw all its pressure on the side bars; then, again, the action will take place at the crown and become so great as to press the bars down in the middle through a distance of many inches, or even to break the stoutest 15-inch oak beams.

This action of the earth, however, seems to be controlled by law, since it depends largely on the depth of the tunnel below the surface. The pressure on the sides is most severe in those parts of the tunnel which are deepest, and the vertical or crown pressure (and this is always the severer of the two) where the distance below ground is less. At first thought this is precisely the reverse of what we should expect to be the case, for it seems but natural tosuppose that the greater the depth of earth the greater the pressure on the arch beneath. The facts are, however, quite the contrary. Thus, for example, in excavating a tunnel through a hill, as we enter the hill side the pressure is almost exclusively at the crown and very severe; as the work progresses nearer and nearer the centre of the hill where the amount of earth above the arch is greatest, the vertical is changed to lateral pressure, and this latter is in turn changed to vertical as we approach the other end. This is well accounted for, by supposing that in the former case the depth of earth being small, the whole of it gets into motion and acts vertically downwards, while in the latter case the amount of earth being great only a small portion is put in motion.

The leading rib, then, must be constructed with no small care, and its joints well strengthened. For tunnels of ordinary span, whatever may be the curve of soffit, we may with safety give the parts the following dimensions. The backpieces two thicknesses of 3 in. plank; the planks breaking joints with each other. For the segment rib make all the braces 6 in. × 6 in.; the long struts reaching from the half sills to the crown 7 in. × 6 in., and the vertical pieces at the crown forming the joint *a b* also 7 in. × 6 in. For the queen post centres, make the tie beam 9 in. × 6 in., as also the short timber placed over the scarf joint; the queen posts 6 in. × 6 in., excepting at the upper and lower ends where the braces abut which should be 10½ in. × 6 in.; the short piece between the queen posts, and just below the crown 4 in. × 6 in., and, finally, the braces 6 in. × 5½ in.

The manner of setting the ribs is illustrated in Figs. 5 and 6. Under the queen post ribs is placed a long horizontal beam, its two ends resting on the side walls and supported immediately under the foot of each queen post by vertical posts. Upon this beam are placed longitudinally four thick planks, and on these rest the folding wedges. The segment ribs are supported in much the same way, each rib by two short timbers, one end of each resting on the side walls and one on a vertical post under the heel of the rib ; on these rest the longitudinal planks which are placed, however, a little oblique to the tunnel since the heel of the segment rib is not so far from the wall as the foot of the queen post.

It has already been remarked that it is never wise to strike the centres until the side walls of the newly excavated length are up, as in work of this class there is a strong tendency to move forward in the direction of the excavation. If, however, the ribs are struck in the manner already described, with the laggings of the back length kept tight up to the arch by the two frames left under them, we shall always have two lengths of completed work remaining with their supports, not only until the next length is excavated but till the side walls are built and ready for the ribs. Under such circumstances each length is well able to uphold its burden till it receives assistance from the next advancing one, the construction of which to springing line occupies several days, and the cement or mortar has time to harden before the weight comes upon the arch after striking the centring. When, however, from false motives of economy, only three ribs and one set of laggings are used, the entire support of one stretch of arch must be removed before another can be commenced, and this, again, before a third is turned, leaving the green arch unsustained, in which state it is liable to give way, the bricks to crush and the whole arch to come down in utter ruin. Nowhere, indeed, among all the variety of engineering works will a penny wise economy more surely prove a pound foolish one than here; nowhere else will an unwise saving lead to so profuse an outlay.

Tunnel centres again differ from those of bridges in that the laggings are very differently adjusted. In the later case it is the custom in practice to place all the laggings on the ribs before commencing to turn the arch, by which means no small degree of stability is given to the ribs. In tunneling, however, where only a few inches of space remains between the backpieces of the frame and the poling which sustains the earth, it would be utterly impossible to turn the arch if *all* the laggings were put in place before the brickwork is begun. To overcome this difficulty, only a few laggings, say five or six are placed at a time. Thus, starting at the springing line, we adjust six laggings on each side of the frame, and carry the arch up equally on both sides. When it has reached the upper bolster, we add six more, and the masonry continued as before, and proceed in this way until very near the crown as shown in Fig. 7, where A A′ is the brick work. At this stage of the work the two laggings C C′ are placed on the ribs, the top of their inner edges being first rabbeted as shown in the figure. In

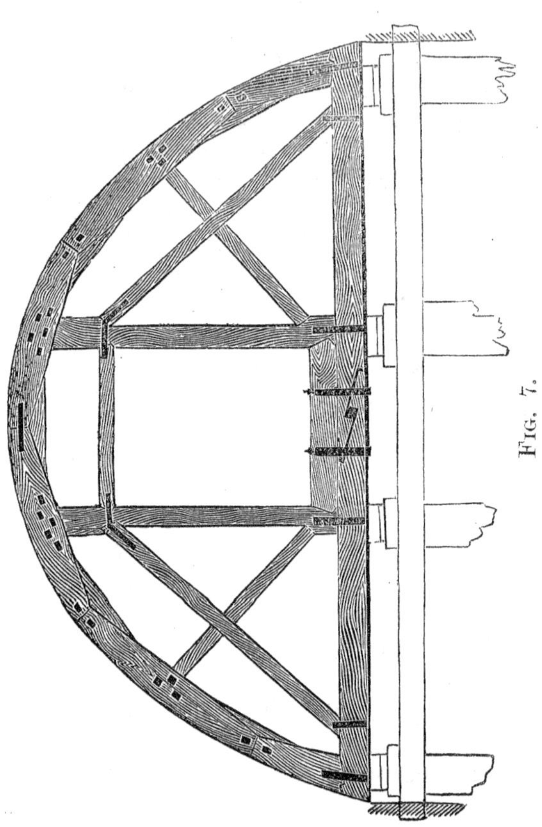

Fig. 7.

these rabbets "*cross*" or "*keying-in*" laggings B, consisting of stout planks 18 or 20 inches in width, are laid one at a time beginning at one end of the centring. The bricklayer whose duty it is to key-in the arch stands with his head and shoulders between the brickwork A, A, and starting at the end of the last piece of completed arch places the first *cross* lagging, and keys in the arch over it; then a second, and in like manner keys in the arch over it, and thus retreats along the entire opening until the whole length of arch is keyed in.

Among the varieties of patent centres that planned by Mr. Frazer, affords a most excellent specimen, and both from its strength, economy, ease of shifting and the small amount of space it occupies in the tunnel, has met with much approval from the engineering profession in England. This centre consists of but three ribs each differing from the other two in design as shown in Figs. 9, 10 and 11, of which 9 is the leading, 10 the middle and 11 the back rib. Each rib is

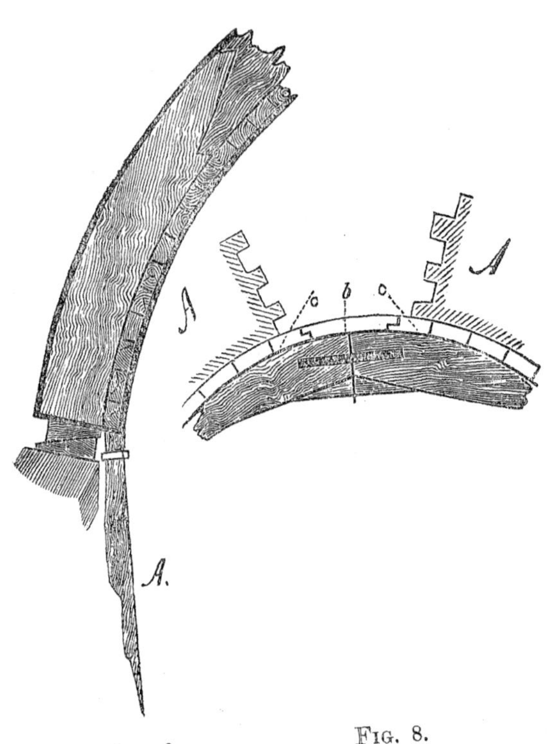

Fig. 9.

Fig. 8.

constructed of four pieces of timber four and one half in. thick by 16 inches wide, scarfed together as shown in the drawings. In centres of the ordinary construction, the ribs when the laggings are laid upon them are all of precisely the same size, and of the same span and rise as the soffit of the intended arch. In Mr. Frazer's plan, however, all the ribs differ in the length of their radii; the radius of the outer curve of the leading rib (Fig. 9) being greater; that of the middle 3 inches less than, and that of the back rib yet smaller than the radius of the soffit; so that the middle centre is the only one of the three which acts in the same way as the ordinary centre frame, that is to say with the laggings and arch resting immediately upon the rib, and is consequently with the laggings on it of the same rise and span as the arch.

The leading rib has for its outlet edge a radius 12½ inches larger than that of the arch soffit, and for its inner edge one 3½ inches less than the same radius (thus making the 16 in. thickness) and is plated on both the inner and outer surface with half inch iron plates bolted quite through. The plate on the inner surface is six inches broad and projects 2 inches over that side of the rib which is turned towards the middle rib, thus forming a flange on which the laggings rest (see Fig. 9). When this rib then is in place, it must be its whole thickness in advance of the end of the intended arch, and as it stands 12½ inches above the soffit will cover 12½ inches of the toothing ends of the brickwork, thus forming a sort of mould to guide the toothing.

The middle rib (Fig. 10) is also covered on the under surface with half inch plate iron in one piece and bolted through as shown in figure, thus giving the rib the strength it would have if supported by the usual struts and braces. The laggings rest immediately upon the upper surface of the rib, and therefore the radius of this side must be the same as that of the arch soffit, less *three inches* to allow for the thickness of the laggings.

Fig. 10.

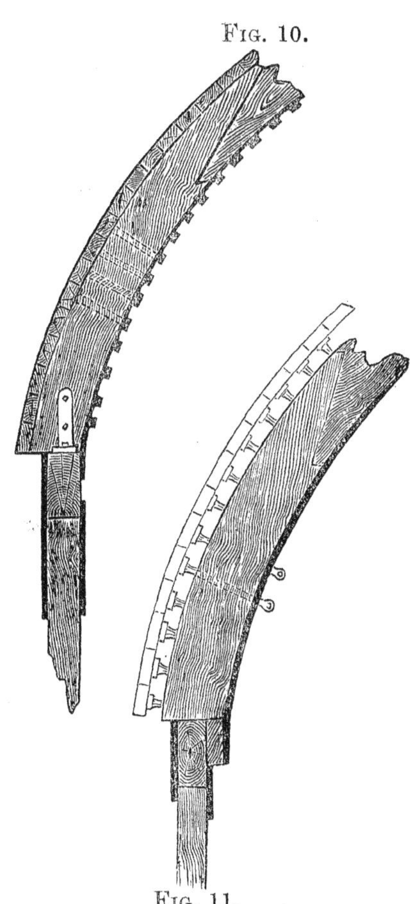

Fig. 11.

The back rib (Fig. 11) is covered on the under surface with a coating of half inch plate iron in one piece, which is bolted through as in the case of the middle rib. Between each bolt a hole is made quite through the rib and its plating, and in it is placed the stem of a bearing iron. There are as many of these irons as there are laggings, the object of using them being to support the laggings which it will be observed do not rest on the rib but on the projecting irons. The amount of projection is regulated by means of adjusting screws, by screwing which the laggings may be raised to the required level, or by unscrewing lowered one by one from the arch when completed. These last two ribs are permanently attached to trestling by brackets, straps and bolts, and the trestling in turn mounted on iron rollers which run on half timbers laid longitudinally as a kind of tramway. They are also steadied at the crown by long iron hooks attached to one rib and fitting into eyes in the other.

The leading rib is supported upon slack blocks placed on top the brick-work of the side walls and by the prop A. This prop, to allow for any inequalities of the invert on which it rests, is mounted at the lower end on a screw by which it may be raised or lowered.

In setting this patent centre, the leading rib is first brought forward into place and wedged up on the edge of the brick-work to its desired level, and the prop A screwed up tight under the heel. The trestles bearing the middle and back ribs are then rolled forward till the middle rib is at the proper distance from the leading one. Three pairs of wedges are then placed between the bottom piece of the trestles and the tramway, and the trestles thus wedged up until the top of the middle rib is on a level with the flange of the leading one, thus giving two level bearings for the laggings. The bearing irons of the back rib are then pushed out by the adjusting screws until the top of each of them is also on a level with the flange of the leading rib. The three bearings then, of each lagging, when the ribs are thus arranged is first upon the flange of the leading rib, then upon the middle rib itself, and finally upon the bearing irons of the back rib. When this centre is to be again moved forward on the completion of this length of arch, a fourth rib called the "*jack rib*" is first fixed under the laggings in the rear of the back rib, this last named rib consists simply of a band of iron 1 inch thick by 2½ wide, bent into the shape of the arch. Opposite every alternate joint of the laggings a screw passes through the rib, and is furnished on its outer end with a square head similar to that of the bearing plates of the back rib, and on its inner or lower end is a loop so that it may be easily turned with a lever. The object of placing these screws opposite each alternate joint is that by this arrangement only half as many screws are needed as there are laggings. The jack rib is itself supported at each end by an iron bar 2 feet long driven temporarily into the wall.

As soon as this latter rib is adjusted to take the ends of the laggings, the wedges are driven from under the trestles and its rollers thus brought down upon the tramway prepared for them. When thus lowered, it is evident that the two ribs (middle and back) will be so much below the leading rib which is left standing that they will easily pass under it. The trestle and its ribs is then moved forward until the back rib is within 8 inches of the ends of the laggings, when it is wedged up as before. The bearing screws are then screwed up tight against the laggings, giving these latter the same support hitherto obtained from the leading rib, which now stands between the middle and back rib. The wedges under the ends of the leading rib (see Fig. 9) are then removed and the rib carried forward over the top of the middle rib and adjusted, as previously described, on the top of the newly built side walls. The laggings are then drawn forward one or two at a time as they are needed, beginning at the springing line.

The great advantage which these patent centres appear to possess over those of the ordinary construction, is the total absence of all struts, ties and braces, thus leaving a fine open space for the scaffolding and materials of the masons. The amount of repairs also is very trivial, as they are not so liable to be injured by flying rocks. In point of economy, though the first cost of patent centres is much greater than that of the segment or queen post centres, the amount expended in repairing the latter soon makes up the difference. In point of strength, it must be acknowledged that, when working through heavy earth, the patent centre of three ribs is by no means so reliable as the all-wood centre of five ribs and two sets of laggings, used as above described. And this is certainly a serious objection in that, it is impossible to tell beforehand at what moment, owing to a fault or to the displacement of the local beds, the character of the earth may change completely from a light soil to one of great heaviness.

*** *Any book in this Catalogue sent free by mail on receipt of price.*

VALUABLE
SCIENTIFIC BOOKS,

PUBLISHED BY

D. VAN NOSTRAND,

23 MURRAY STREET AND 27 WARREN STREET,

NEW YORK.

FRANCIS. Lowell Hydraulic Experiments, being a selection from Experiments on Hydraulic Motors, on the Flow of Water over Weirs, in Open Canals of Uniform Rectangular Section, and through submerged Orifices and diverging Tubes. Made at Lowell, Massachusetts. By James B. Francis, C. E. 2d edition, revised and enlarged, with many new experiments, and illustrated with twenty-three copperplate engravings. 1 vol. 4to, cloth......................$15 00

ROEBLING (J. A.) Long and Short Span Railway Bridges. By John A. Roebling, C. E. Illustrated with large copperplate engravings of plans and views. Imperial folio, cloth.............................. 25 00

CLARKE (T. C.) Description of the Iron Railway Bridge over the Mississippi River, at Quincy, Illinois. Thomas Curtis Clarke, Chief Engineer. Illustrated with 21 lithographed plans. 1 vol. 4to, cloth.. 7 50

TUNNER (P.) A Treatise on Roll-Turning for the Manufacture of Iron. By Peter Tunner. Translated and adapted by John B. Pearse, of the Pennylvania Steel Works, with numerous engravings wood cuts and folio atlas of plates.................. $10 00

ISHERWOOD (B. F.) Engineering Precedents for Steam Machinery. Arranged in the most practical and useful manner for Engineers. By B. F. Isherwood, Civil Engineer, U. S. Navy. With Illustrations. Two volumes in one. 8vo, cloth.......... $2 50

BAUERMAN. Treatise on the Metallurgy of Iron, containing outlines of the History of Iron Manufacture, methods of Assay, and analysis of Iron Ores, processes of manufacture of Iron and Steel, etc., etc. By H. Bauerman. First American edition. Revised and enlarged, with an Appendix on the Martin Process for making Steel, from the report of Abram S. Hewitt. Illustrated with numerous wood engravings. 12mo, cloth.................................... 2 00

CAMPIN on the Construction of Iron Roofs. By Francis Campin. 8vo, with plates, cloth.......... 2 00

COLLINS. The Private Book of Useful Alloys and Memoranda for Goldsmiths, Jewellers, &c. By James E. Collins. 18mo, cloth.................. 75

CIPHER AND SECRET LETTER AND TELEGRAPHIC CODE, with Hogg's Improvements. The most perfect secret code ever invented or discovered. Impossible to read without the key. By C. S. Larrabee. 18mo, cloth..................... 1 00

COLBURN. The Gas Works of London. By Zerah Colburn, C. E. 1 vol 12mo, boards.............. 60

CRAIG (B. F.) Weights and Measures. An account of the Decimal System, with Tables of Conversion for Commercial and Scientific Uses. By B. F. Craig, M.D. 1 vol. square 32mo, limp cloth.... 50

NUGENT. Treatise on Optics; or, Light and Sight, theoretically and practically treated; with the application to Fine Art and Industrial Pursuits. By E. Nugent. With one hundred and three illustrations. 12mo, cloth.. 2 00

FREE HAND DRAWING. A Guide to Ornamental Figure and Landscape Drawing. By an Art Student. 18mo, boards,............................ 50

HOWARD. Earthwork Mensuration on the Basis of the Prismoidal Formulae. Containing simple and labor-saving method of obtaining Prismoidal contents directly from End Areas. Illustrated by Examples, and accompanied by Plain Rules for Practical Uses. By Conway R. Howard, C. E., Richmond, Va. Illustrated, 8vo, cloth............................ 1 50

GRUNER. The Manufacture of Steel. By M. L. Gruner. Translated from the French, by Lenox Smith, with an appendix on the Bessamer process in the United States, by the translator. Illustrated by Lithographed drawings and wood cuts. 8vo, cloth.. 3 50

AUCHINCLOSS. Link and Valve Motions Simplified. Illustrated with 37 wood-cuts, and 21 lithographic plates, together with a Travel Scale, and numerous useful Tables. By W. S. Auchincloss. 8vo, cloth.. 3 00

VAN BUREN. Investigations of Formulas, for the strength of the Iron parts of Steam Machinery. By J. D. Van Buren, Jr., C. E. Illustrated, 8vo, cloth. 2 00

JOYNSON. Designing and Construction of Machine Gearing. Illustrated, 8vo, cloth.................. 2 00

GILLMORE. Coignet Beton and other Artificial Stone. By Q. A. Gillmore, Major U. S. Corps Engineers. 9 plates, views, &c. 8vo, cloth.................. 2 50

SAELTZER. Treattse on Acoustics in connection with Ventilation. By Alexander Saeltzer, Architect. 12mo, cloth.............................. 2 00

THE EARTH'S CRUST. A handy Outline of Geology. By David Page. Illustrated, 18mo, cloth.... 75

DICTIONARY of Manufactures, Mining, Machinery, and the Industrial Arts. By George Dodd. 12mo, cloth.. 2 00

BOW. A Treatise on Bracing, with its application to Bridges and other Structures of Wood or Iron. By Robert Henry Bow, C. E. 156 illustrations, 8vo, cloth.................................... 1 50

GILLMORE (Gen. Q. A.) Treatise on Limes, Hydraulic Cements, and Mortars. Papers on Practical Engineering, U. S. Engineer Department, No. 9, containing Reports of numerous Experiments conducted in New York City, during the years 1858 to 1861, inclusive. By Q. A. Gillmore, Bvt. Maj -Gen., U. S. A., Major, Corps of Engineers. With numerous illustrations. 1 vol, 8vo, cloth............... $4 00

HARRISON. The Mechanic's Tool Book, with Practical Rules and Suggestions for Use of Machinists, Iron Workers, and others. By W. B. Harrison, associate editor of the "American Artisan." Illustrated with 44 engravings. 12mo, cloth............ 1 50

HENRICI (Olaus). Skeleton Structures, especially in their application to the Building of Steel and Iron Bridges. By Olaus Henrici. With folding plates and diagrams. 1 vol. 8vo, cloth.................... 1 50

HEWSON (Wm.) Principles and Practice of Embanking Lands from River Floods, as applied to the Levees of the Mississippi. By William Hewson, Civil Engineer. 1 vol. 8vo, cloth...................... 2 00

HOLLEY (A. L.) Railway Practice. American and European Railway Practice, in the economical Generation of Steam, including the Materials and Construction of Coal-burning Boilers, Combustion, the Variable Blast, Vaporization, Circulation, Superheating, Supplying and Heating Feed-water, etc., and the Adaptation of Wood and Coke-burning Engines to Coal-burning; and in Permanent Way, including Road-bed, Sleepers, Rails, Joint-fastenings, Street Railways, etc., etc. By Alexander L. Holley, B. P. With 77 lithographed plates. 1 vol. folio, cloth.... 12 00

KING (W. H.) Lessons and Practical Notes on Steam, the Steam Engine, Propellers, etc., etc., for Young Marine Engineers, Students, and others. By the late W. H. King, U. S. Navy. Revised by Chief Engineer J. W. King, U. S. Navy. Twelfth edition, enlarged. 8vo, cloth. 2 00

MINIFIE (Wm.) Mechanical Drawing. A Text-Book of Geometrical Drawing for the use of Mechanics and Schools, in which the Definitions and Rules of Geometry are familiarly explained; the Practical Problems are arranged, from the most simple to the more complex, and in their description technicalities are avoided as much as possible. With illustrations for Drawing Plans, Sections, and Elevations of Railways and Machinery; an Introduction to Isometrical Drawing, and an Essay on Linear Perspective and Shadows. Illustrated with over 200 diagrams engraved on steel. By Wm. Minifie, Architect. Seventh edition. With an Appendix on the Theory and Application of Colors. 1 vol. 8vo, cloth........... $4 00

"It is the best work on Drawing that we have ever seen, and is especially a text-book of Geometrical Drawing for the use of Mechanics and Schools. No young Mechanic, such as a Machinists, Engineer, Cabinet-maker, Millwright, or Carpenter, should be without it."—*Scientific American.*

—— Geometrical Drawing. Abridged from the octavo edition, for the use of Schools. Illustrated with 48 steel plates. Fifth edition. 1 vol. 12mo, cloth.... 2 00

STILLMAN (Paul.) Steam Engine Indicator, and the Improved Manometer Steam and Vacuum Gauges—their Utility and Application. By Paul Stillman. New edition. 1 vol. 12mo, flexible cloth........... 1 00

SWEET (S. H.) Special Report on Coal; showing its Distribution, Classification, and cost delivered over different routes to various points in the State of New York, and the principal cities on the Atlantic Coast. By S. H. Sweet. With maps, 1 vol. 8vo, cloth..... 3 00

WALKER (W. H.) Screw Propulsion. Notes on Screw Propulsion: its Rise and History. By Capt. W. H. Walker, U. S. Navy. 1 vol. 8vo, cloth..... 75

WARD (J. H.) Steam for the Million. A popular Treatise on Steam and its Application to the Useful Arts, especially to Navigation. By J. H. Ward, Commander U. S. Navy. New and revised edition. 1 vol. 8vo, cloth................................. 1 00

WEISBACH (Julius). Principles of the Mechanics of Machinery and Engineering. By Dr. Julius Weisbach, of Freiburg. Translated from the last German edition. Vol. I., 8vo, cloth.. .... 10 00

DIEDRICH. The Theory of Strains, a Compendium for the calculation and construction of Bridges, Roofs, and Cranes, with the application of Trigonometrical Notes, containing the most comprehensive information in regard to the Resulting strains for a permanent Load, as also for a combined (Permanent and Rolling) Load. In two sections, adadted to the requirements of the present time. By John Diedrich, C. E. Illustrated by numerous plates and diagrams. 8vo, cloth.................................. 5 00

WILLIAMSON (R. S.) On the use of the Barometer on Surveys and Reconnoissances. Part I. Meteorology in its Connection with Hypsometry. Part II. Barometric Hypsometry. By R. S. Wiliamson, Bvt. Lieut.-Col. U. S. A., Major Corps of Engineers. With Illustrative Tables and Engravings. Paper No. 15, Professional Papers, Corps of Engineers. 1 vol. 4to, cloth.................................. 15 00

POOK (S. M.) Method of Comparing the Lines and Draughting Vessels Propelled by Sail or Steam. Including a chapter on Laying off on the Mould-Loft Floor. By Samuel M. Pook, Naval Constructor. 1 vol. 8vo, with illustrations, cloth............ 5 00

ALEXANDER (J. H.) Universal Dictionary of Weights and Measures, Ancient and Modern, reduced to the standards of the United States of America. By J. H. Alexander. New edition, enlarged. 1 vol. 8vo, cloth.................................. 3 50

WANKLYN. A Practical Treatise on the Examination of Milk, and its Derivatives, Cream, Butter and Cheese. By J. Alfred Wanklyn, M. R. C. S., 12mo, cloth.................................. 1 00

RICHARDS' INDICATOR. A Treatise on the Richards Steam Engine Indicator, with an Appendix by F. W. Bacon, M. E. 18mo, flexible, cloth.......... 1 00

POPE. Modern Practice of the Electric Telegraph. A Hand Book for Electricians and operators. By Frank L. Pope. Eighth edition, revised and enlarged, and fully illlustrated. 8vo, cloth........................ $2.00

"There is no other work of this kind in the English language that contains in so small a compass so much practical information in the application of galvanic electricity to telegraphy. It should be in the hands of every one interested in telegraphy, or the use of Batteries for other purposes."

MORSE. Examination of the Telegraphic Apparatus and the Processes in Telegraphy. By Samuel F. Morse, LL.D., U. S. Commissioner Paris Universal Exposition, 1867. Illustrated, 8vo, cloth............ $2 00

SABINE. History and Progress of the Electric Telegraph, with descriptions of some of the apparatus. By Robert Sabine, C. E. Second edition, with additions, 12mo, cloth.................................. 1 25

BLAKE. Ceramic Art. A Report on Pottery, Porcelain, Tiles, Terra Cotta and Brick. By W. P. Blake, U. S. Commissioner, Vienna Exhibition, 1873. 8vo, cloth.. 2 00

BENET. Electro-Ballistic Machines, and the Schultz Chronoscope. By Lieut.-Col. S. V. Benet, Captain of Ordnance, U. S. Army. Illustrated, second edition, 4to, cloth...................................... 3 00

MICHAELIS. The Le Boulenge Chronograph, with three Lithograph folding plates of illustrations. By Brevet Captain O. E. Michaelis, First Lieutenant Ordnance Corps, U. S. Army, . 4to, cloth.......... 3 00

ENGINEERING FACTS AND FIGURES An Annual Register of Progress in Mechanical Engineering and Construction, for the years 1863, 64, 65, 66, 67, 68. Fully illustrated, 6 vols. 18mo, cloth, $2.50 per vol., each volume sold separately..............

HAMILTON. Useful Information for Railway Men. Compiled by W. G. Hamilton, Engineer. Fifth edition, revised and enlarged, 562 pages Pocket form. Morocco, gilt.. 2 00

STUART. The Civil and Military Engineers of America. By Gen. C. B. Stuart. With 9 finely executed portraits of eminent engineers, and illustrated by engravings of some of the most important works constructed in America. 8vo, cloth.................. $5 00

STONEY. The Theory of Strains in Girders and similar structures, with observations on the application of Theory to Practice, and Tables of Strength and other properties of Materials. By Bindon B. Stoney, B. A. New and revised edition, enlarged, with numerous engravings on wood, by Oldham. Royal 8vo, 664 pages. Complete in one volume. 8vo, cloth....... 12 50

SHREVE. A Treatise on the Strength of Bridges and Roofs. Comprising the determination of Algebraic formulas for strains in Horizontal, Inclined or Rafter, Triangular, Bowstring, Lenticular and other Trusses, from fixed and moving loads, with practical applications and examples, for the use of Students and Engineers. By Samuel H. Shreve, A. M., Civil Engineer. 87 wood cut illustrations. 8vo, cloth............... 5 00

MERRILL. Iron Truss Bridges for Railroads. The method of calculating strains in Trusses, with a careful comparison of the most prominent Trusses, in reference to economy in combination, etc., etc. By Brevet. Col. William E. Merrill, U. S. A., Major Corps of Engineers, with nine lithographed plates of Illustrations. 4to, cloth........................... 5 00

WHIPPLE. An Elementary and Practical Treatise on Bridge Building. An enlarged and improved edition of the author's original work. By S. Whipple, C. E., inventor of the Whipple Bridges, &c. Illustrated 8vo, cloth....................................... 4 00

THE KANSAS CITY BRIDGE. With an account of the Regimen of the Missouri River, and a description of the methods used for Founding in that River. By O Chanute, Chief Engineer, and George Morrison, Assistant Engineer. Illustrated with five lithographic views and twelve plates of plans. 4to, cloth, 6 00

MAC CORD. A Practical Treatise on the Slide Valve by Eccentrics, examining by methods the action of the Eccentric upon the Slide Valve, and explaining the Practical processes of laying out the movements, adapting the valve for its various duties in the steam engine. For the use of Engineers, Draughtsmen, Machinists, and Students of Valve Motions in general. By C. W. Mac Cord, A. M., Professor of Mechanical Drawing, Stevens' Institute of Technology, Hoboken, N. J. Illustrated by 8 full page copper-plates. 4to, cloth.......... $4 00

KIRKWOOD. Report on the Filtration of River Waters, for the supply of cities, as practised in Europe, made to the Board of Water Commissioners of the City of St. Louis. By James P. Kirkwood. Illustrated by 30 double plate engravings. 4to, cloth, 15 00

PLATTNER. Manual of Qualitative and Quantitative Analysis with the Blow Pipe. From the last German edition, revised and enlarged. By Prof. Th. Richter, of the Royal Saxon Mining Academy. Translated by Prof. H. B. Cornwall, Assistant in the Columbia School of Mines, New York assisted by John H. Caswell. Illustrated with 87 wood cuts, and one lithographic plate. Second edition, revised, 560 pages, 8vo, cloth.......... 7 50

PLYMPTON. The Blow Pipe. A Guide to its Use in the Determination of Salts and Minerals. Compiled from various sources, by George W. Plympton, C. E. A. M., Professor of Physical Science in the Polytechnic Institute, Brooklyn, New York, 12mo, cloth.......... 1 50

PYNCHON. Introduction to Chemical Physics, designed for the use of Academies, Colleges and High Schools. Illustrated with numerous engravings, and containing copious experiments with directions for preparing them. By Thomas Ruggles Pynchon, M. A., Professor of Chemistry and the Natural Sciences, Trinity College, Hartford New edition, revised and enlarged, and illustrated by 269 illustrations on wood. Crown, 8vo, cloth.......... 3 00

ELIOT AND STORER. A compendious Manual of Qualitative Chemical Analysis. By Charles W. Eliot and Frank H. Storer. Revised with the Co-operation of the authors. By William R. Nichols, Professor of Chemistry in the Massachusetts Institute of Technology. Illustrated, 12mo, cloth....... $1 50

RAMMELSBERG. Guide to a course of Quantitative Chemical Analysis, especially of Minerals and Furnace Products. Illustrated by Examples. By C. F. Rammelsberg. Translated by J. Towler, M. D. 8vo, cloth.. 2 25

EGLESTON. Lectures on Descriptive Mineralogy, delivered at the School of Mines, Columbia College. By Professor T. Egleston. Illustrated by 34 Lithographic Plates. 8vo, cloth........................ 4 50

JACOB. On the Designing and Construction of Storage Reservoirs, with Tables and Wood Cuts representing Sections, &c., 18mo, boards..................... 50

WATT'S Dictionary of Chemistry. New and Revised edition complete in 6 vols. 8vo. cloth, $62.00. Supplementary volume sold separately. Price, cloth... 9 00

RANDALL. Quartz Operators Hand-Book. By P. M. Randall. New edition, revised and enlarged, fully illustrated. 12mo, cloth 2 00

SILVERSMITH. A Practical Hand-Book for Miners, Metallurgists, and Assayers, comprising the most recent improvements in the disintegration, amalgamation, smelting, and parting of the Precious ores, with a comprehensive Digest of the Mining Laws. Greatly augmented, revised and corrected. By Julius Silversmith. Fourth edition. Profusely illustrated. 12mo, cloth.. 3 00

THE USEFUL METALS AND THEIR ALLOYS, including Mining Ventilation, Mining Jurisprudence, and Metallurgic Chemistry employed in the conversion of Iron, Copper, Tin, Zinc, Antimony and Lead ores, with their applications to the Industrial Arts. By Scoffren, Truan, Clay, Oxland, Fairbairn, and others. Fifth edition, half calf.................. 3 75

JOYNSON. The Metals used in construction, Iron, Steel, Bessemer Metal, etc., etc. By F. H. Joynson. Illustrated, 12mo, cloth.......................... $0 75

VON COTTA. Treatise on Ore Deposits. By Bernhard Von Cotta, Professor of Geology in the Royal School of Mines, Freidberg, Saxony. Translated from the second German edition, by Frederick Prime, Jr., Mining Engineer, and revised by the author, with numerous illustrations. 8vo, cloth....... 4 00

GREENE. Graphical Method for the Analysis of Bridge Trusses, extended to continuous Girders and Draw Spans. By C. E. Greene, A. M., Prof. of Civil Engineering, University of Michigan. Illustrated by 3 folding plates, 8vo, cloth.......................... 2 00

BELL. Chemical Phenomena of Iron Smelting. An experimental and practical examination of the circumstances which determine the capacity of the Blast Furnace, The Temperature of the air, and the proper condition of the Materials to be operated upon. By I. Lowthian Bell. 8vo, cloth........... 6 00

ROGERS. The Geology of Pennsylvania. A Government survey, with a general view of the Geology of the United States, Essays on the Coal Formation and its Fossils, and a description of the Coal Fields of North America and Great Britain. By Henry Darwin Rogers, late State Geologist of Pennsylvania, Splendidly illustrated with Plates and Engravings in the text. 3 vols., 4to, cloth, with Portfolio of Maps. 30 00

BURGH. Modern Marine Engineering, applied to Paddle and Screw Propulsion. Consisting of 36 colored plates, 259 Practical Wood Cut Illustrations, and 403 pages of descriptive matter, the whole being an exposition of the present practice of James Watt & Co., J. & G. Rennie, R. Napier & Sons, and other celebrated firms, by N. P. Burgh, Engineer, thick 4to, vol., cloth, $25.00; half mor....... 30 00

CHURCH. Notes of a Metallurgical Journey in Europe. By J. A. Church, Engineer of Mines, 8vo, cloth..... 2 00

BOURNE. Treatise on the Steam Engine in its various applications to Mines, Mills, Steam Navigation, Railways, and Agriculture, with the theoretical investigations respecting the Motive Power of Heat, and the proper proportions of steam engines. Elaborate tables of the right dimensions of every part, and Practical Instructions for the manufacture and management of every species of Engine in actual use. By John Bourne, being the ninth edition of "A Treatise on the Steam Engine," by the "Artizan Club." Illustrated by 38 plates and 546 wood cuts. 4to, cloth.................................... $15 00

STUART. The Naval Dry Docks of the United States. By Charles B. Stuart late Engineer-in-Chief of the U. S. Navy. Illustrated with 24 engravings on steel. Fourth edition, cloth.................... 6 00

ATKINSON. Practical Treatises on the Gases met with in Coal Mines. 18mo, boards................ 50

FOSTER. Submarine Blasting in Boston Harbor, Massachusetts. Removal of Tower and Corwin Rocks. By J. G. Foster, Lieut.-Col. of Engineers, U. S. Army. Illustrated with seven plates, 4to, cloth... 3 50

BARNES Submarine Warfare, offensive and defensive, including a discussion of the offensive Torpedo System, its effects upon Iron Clad Ship Systems and influence upon future naval wars. By Lieut.-Commander J. S. Barnes, U. S. N., with twenty lithographic plates and many wood cuts. 8vo, cloth..... 5 00

HOLLEY. A Treatise on Ordnance and Armor, embracing descriptions, discussions, and professional opinions concerning the materials, fabrication, requirements, capabilities, and endurance of European and American Guns, for Naval, Sea Coast, and Iron Clad Warfare, and their Rifling, Projectiles, and Breech-Loading; also, results of experiments against armor, from official records, with an appendix referring to Gun Cotton, Hooped Guns, etc., etc. By Alexander L. Holley, B. P., 948 pages, 493 engravings, and 147 Tables of Results, etc., 8vo, half roan. 10 00

SIMMS. A Treatise on the Principles and Practice of Levelling, showing its application to purposes of Railway Engineering and the Construction of Roads, &c. By Frederick W. Simms, C. E. From the 5th London edition, revised and corrected, with the addition of Mr. Laws's Practical Examples for setting out Railway Curves. Illustrated with three Lithographic plates and numerous wood cuts. 8vo, cloth. $2 50

BURT. Key to the Solar Compass, and Surveyor's Companion; comprising all the rules necessary for use in the field; also description of the Linear Surveys and Public Land System of the United States, Notes on the Barometer, suggestions for an outfit for a survey of four months, etc. By W. A. Burt, U. S. Deputy Surveyor. Second edition. Pocket book form, tuck.. 2 50

THE PLANE TABLE. Its uses in Topographical Surveying, from the Papers of the U. S. Coast Survey. Illustrated, 8vo, cloth........................ 2 00

"This work gives a description of the Plane Table, employed at the U. S. Coast Survey office, and the manner of using it."

JEFFER'S. Nautical Surveying. By W. N. Jeffers, Captain U. S. Navy. Illustrated with 9 copperplates and 31 wood cut illustrations. 8vo, cloth........... 5 00

CHAUVENET. New method of correcting Lunar Distances, and improved method of Finding the error and rate of a chronometer, by equal altitudes. By W. Chauvenet, LL.D. 8vo, cloth................. 2 00

BRUNNOW. Spherical Astronomy. By F. Brunnow, Ph. Dr. Translated by the author from the second German edition. 8vo, cloth...................... 6 50

PEIRCE. System of Analytic Mechanics. By Benjamin Peirce. 4to, cloth.......................... 10 00

COFFIN. Navigation and Nautical Astronomy. Prepared for the use of the U. S. Naval Academy. By Prof. J. H. C. Coffin. Fifth edition. 52 wood cut illustrations. 12mo, cloth.......................... 3 50

CLARK. Theoretical Navigation and Nautical Astronomy. By Lieut. Lewis Clark, U. S. N. Illustrated with 41 wood cuts. 8vo, cloth.................... $3 00

HASKINS. The Galvanometer and its Uses. A Manual for Electricians and Students. By C. H. Haskins. 12mo, pocket form, morocco. (In press).....

GOUGE. New System of Ventilation, which has been thoroughly tested, under the patronage of many distinguished persons. By Henry A. Gouge. With many illustrations. 8vo, cloth.................... 2 00

BECKWITH. Observations on the Materials and Manufacture of Terra-Cotta, Stone Ware, Fire Brick, Porcelain and Encaustic Tiles, with remarks on the products exhibited at the London International Exhibition, 1871. By Arthur Beckwith, C. E. 8vo, paper.................... 60

MORFIT. A Practical Treatise on Pure Fertilizers, and the chemical conversion of Rock Guano, Marlstones, Coprolites, and the Crude Phosphates of Lime and Alumina generally, into various valuable products. By Campbell Morfit, M.D., with 28 illustrative plates, 8vo, cloth.................... 20 00

BARNARD. The Metric System of Weights and Measures. An address delivered before the convocation of the University of the State of New York, at Albany, August, 1871. By F. A. P. Barnard, LL.D., President of Columbia College, New York. Second edition from the revised edition, printed for the Trustees of Columbia College. Tinted paper, 8vo, cloth 3 00

—— Report on Machinery and Processes on the Industrial Arts and Apparatus of the Exact Sciences. By F. A. P. Barnard, LL.D. Paris Universal Exposition, 1867. Illustrated, 8vo, cloth.............. 5 00

ALLAN. Theory of Arches. By Prof. W. Allan, formerly of Washington & Lee University, 18mo, b'rds 50

MYER. Manual of Signals, for the use of Signal officers in the Field, and for Military and Naval Students, Military Schools, etc. A new edition enlarged and illustrated By Brig. General Albert J. Myer, Chief Signal Officer of the army, Colonel of the Signal Corps during the War of the Rebellion. 12mo, 48 plates, full Roan.. $5 00

WILLIAMSON. Practical Tables in Meteorology and Hypsometry, in connection with the use of the Barometer. By Col. R. S. Williamson, U. S. A. 4to, cloth.. 2 50

CLEVENGER. A Treatise on the Method of Government Surveying, as prescribed by the U. S. Congress and Commissioner of the General Land Office, with complete Mathematical, Astronomical and Practical Instructions for the Use of the United States Surveyors in the Field. By S. R. Clevenger, Pocket Book Form, Morocco.. 2 50

PICKERT AND METCALF. The Art of Graining. How Acquired and How Produced, with description of colors, and their application. By Charles Pickert and Abraham Metcalf. Beautifully illustrated with 42 tinted plates of the various woods used in interior finishing. Tinted paper, 4to, cloth.. 10 00

HUNT. Designs for the Gateways of the Southern Entrances to the Central Park. By Richard M. Hunt. With a description of the designs. 4to. cloth...... 5 00

LAZELLE. One Law in Nature. By Capt. H. M. Lazelle, U. S. A. A new Corpuscular Theory, comprehending Unity of Force, Identity of Matter, and its Multiple Atom Constitution, applied to the Physical Affections or Modes of Energy. 12mo, cloth... 1 50

CORFIELD. Water and Water Supply. By W. H. Corfield, M. A. M, D., Professor of Hygiene and Public Health at University College, London. 18mo, boards.. 50

BOYNTON. History of West Point, its Military Importance during the American Revolution, and the Origin and History of the U. S. Military Academy. By Bvt. Major C. E. Boynton, A.M., Adjutant of the Military Academy. Second edition, 416 pp. 8vo, printed on tinted paper, beautifully illustrated with 36 maps and fine engravings, chiefly from photographs taken on the spot by the author. Extra cloth.. $3 50

WOOD. West Point Scrap Book, being a collection of Legends, Stories, Songs, etc., of the U. S. Military Academy. By Lieut. O. E. Wood, U. S. A. Illustrated by 69 engravings and a copperplate map. Beautifully printed on tinted paper. 8vo, cloth..... 5 00

WEST POINT LIFE. A Poem read before the Dialectic Society of the United States Military Academy. Illustrated with Pen-and-Ink Sketches. By a Cadet. To which is added the song, "Benny Havens, oh!" oblong 8vo, 21 full page illustrations, cloth.......... 2 50

GUIDE TO WEST POINT and the U. S. Military Academy, with maps and engravings, 18mo, blue cloth, flexible.. 1 00

HENRY. Military Record of Civilian Appointments in the United States Army. By Guy V. Henry, Brevet Colonel and Captain First United States Artillery, Late Colonel and Brevet Brigadier General, United States Volunteers. Vol. 1 now ready. Vol. 2 in press. 8vo, per volume, cloth.......................... 5 00

HAMERSLY. Records of Living Officers of the U. S. Navy and Marine Corps. Compiled from official sources. By Lewis B. Hamersly, late Lieutenant U. S. Marine Corps. Revised edition, 8vo, cloth... 5 00

MOORE. Portrait Gallery of the War. Civil, Military and Naval. A Biographical record, edited by Frank Moore. 60 fine portraits on steel. Royal 8vo, cloth.. 6 00

PRESCOTT. Outlines of Proximate Organic Analysis, for the Identification, Separation, and Quantitative Determination of the more commonly occurring Organic Compounds. By Albert B. Prescott, Professor of Chemistry, University of Michigan, 12mo, cloth... 1 75

PRESCOTT. Chemical Examination of Alcoholic Liquors. A Manual of the Constituents of the Distilled Spirits and Fermented Liquors of Commerce, and their Qualitative and Quantitative Determinations. By Albert B. Prescott, 12mo, cloth.................. 1 50

BATTERSHALL. Legal Chemistry. A Guide to the Detection of Poisons, Falsification of Writings, Adulteration of Alimentary and Pharmaceutical Substances; Analysis of Ashes, and examination of Hair, Coins, Arms and Stains, as applied to Chemical Jurisprudence, for the Use of Chemists, Physicians, Lawyers, Pharmacists and Experts Translated with additions, including a list of books and Memoirs on Texicology, etc. from the French of A. Naquet. By J. P. Battershall, Ph. D. with a preface by C. F. Chandler, Ph. D., M. D,, L. L. D. 12mo, cloth.... —

McCULLOCH. Elementary Treatise on the Mechanical Theory of Heat, and its application to Air and Steam Engines. By R. S. McCulloch, 8vo, cloth.... —

AXON. The Mechanics Friend; a Collection of Receipts and Practical Suggestions Relating to Aquaria—Bronzing—Cements—Drawing—Dyes— Electricity—Gilding—Glass Working—Glues—Horology—Lacquers—Locomotives—Magnetism—Metal-Working—Modelling—Photography—Pyrotechy—Railways—Solders—Steam Engine—Telegraphy—Taxidermy—Varnishes — Water-Proofing and Miscellaneous Tools,—Instruments, Machines and Processes connected with the Chemical and Mechanics Arts; with numerous diagrams and wood cuts. Edited by William E. A. Axon. Fancy cloth..................... 1 50

ERNST. Manual of Practical Military Engineering, Prepared for the use of the Cadets of the U. S. Military Academy, and for Engineer Troops. By Capt. O. H. Ernst, Corps of Engineers, Instructor in Practical Military Engineering, U. S. Military Academy. 192 wood cuts and 3 lithographed plates. 12mo, cloth.. 5 00

BUTLER. Projectiles and Rifled Cannon. A Critical Discussion of the Principal Systems of Rifling and Projectiles, with Practical Suggestions for their Improvement, as embraced in a Report to the Chief of Ordnance, U. S. A. By Capt. John S. Butler, Ordnance Corps, U. S. A. 36 plates, 4to, cloth........ 7 50

BLAKE. Report upon the Precious Metals: Being Statistical Notices of the principal Gold and Silver producing regions of the World, Represented at the Paris Universal Exposition. By William P. Blake, Commissioner from the State of California. 8vo, cloth 2 00

TONER. Dictionary of Elevations and Climatic Register of the United States. Containing in addition to Elevations, the Latitude, Mean, Annual Temperature, and the total Annual Rain fall of many localities; with a brief introduction on the Orographic and Physical Peculiarities of North America. By J. M. Toner, M. D. 8vo, cloth.............................. 3 75

MOWBRAY. Tri-Nitro Glycerine, as applied in the Hoosac Tunnel, and to Submarine Blasting, Torpedoes, Quarrying, etc. Being the result of six year's observation and practice during the manufacture of five hundred thousand pounds of this explosive Mica, Blasting Powder, Dynamites; with an account of the various Systems of Blasting by Electricity, Priming Compounds, Explosives, etc., etc. By George M. Mowbray, Operative Chemist, with thirteen illustrations, tables and appendix. Third Edition. Re-written. 8vo. cloth.............................. 3 00

www.ingramcontent.com/pod-product-compliance
Lightning Source LLC
LaVergne TN
LVHW021421110826
845150LV00007B/2021